I0393808

Table of Contents

List of Figures

List of Tables

Augmenting Latent Dirichlet Allocation

and

Rank Threshold Detection with Ontologies

I. Introduction

The usefulness of information often depends on the efficient extraction of relevant information. In the area of intelligence analysis, data management systems often become overwhelmed with source documents, in free text form, that are not labeled or pre-assigned to specific topics. Automatic document modeling, document classification and topic categorization algorithms are used to help solve this problem [7]. One specific technique, Latent Dirichlet Allocation (LDA) [7], is a generative model that assigns documents to discovered topics and words (or terms) to topics with some probability. However, due to English words having synonyms, this type of probabilistic clustering can be misled, often resulting in misclassification of words to topics.

To help resolve the true semantics of a word from a document point of view, its context must be taken into account. Moreover, other document information can be saved, such as neighboring terms, to help determine its context [7]. However, to improve performance, most automated systems assume word independence and use a unigram approach, i.e., documents are assumed to be composed of an unrelated "bag of words," which relieve systems from maintaining a combinatorial representation of related words. However, losing this context sacrifices the system's ability to conduct a more correct semantic analysis of each word in each document.

One way to maintain word semantic relationships is to develop a persistent semantic ontology to maintain groups of concepts and their relationships among other concepts. One such worldwide open-source project is called the WordNet ontology [46]. This thesis shows how ontologies can be used to augment and improve document modeling algorithms. Specifically, we investigate the benefits of incorpo-

1

rating WordNet into the LDA algorithm. LDA is a probabilistic topic model that can infer a topic distribution based on word content for each document in a corpus. This inference capability is extended by highlighting semantic relationships that may be concealed, i.e., having low word probability, and thereby discounted by LDA. Favorable results conclude that the LDA with WordNet (LDAWN) algorithm generated lower perplexity results over LDA alone suggesting that ontology augmentation is beneficial for document modeling refinement.

LDAWN also incorporates a query model for information retrieval purposes. The returned documents are ranked according to their relevance to a given query by combining the Dirichlet smoothing document model with the LDA model as proposed by Wei and Croft [48]. Previous work by Millar [36] combined LDA with Self-Organizing Maps (LDA-SOM), which rank document relevance to a query independent of whether or not the query terms appeared in the document. More importantly, Wei and Croft's model and LDA-SOM overlook query terms that do not explicitly co-occur and are discounted by LDA. In LDAWN, WordNet is used to rank documents based on the query terms including any of its synsets to leverage terms that co-occur. Furthermore, LDAWN is used to automatically locate and label the relevancy threshold in the ranked documents.

The LDAWN process exposes hidden semantic relationships resulting in improved document modeling, document classification and topic categorization over LDA alone. For any given document, term frequencies are incremented for all terms in the document with matching terms in WordNet synsets. Then, the resulting term-document matrix is incorporated into the LDA model to compute the topic distribution. LDA estimates the per-document topic distribution and per-topic word distribution and outputs the probabilities for each topic distribution. Then we compare the performance of LDAWN against LDA alone by training 90% for each of four corpora. After this unigram classification training, the held-out test set is used to measure the perplexity of each collection over several numbers of topics. These tests were repeated over five randomized versions of each corpus. Our results show that LDAWN achieved lower perplexity values than basic LDA, i.e., LDAWN provides a better generalization of unseen documents.

II. Background

Tools used for document modeling face difficult challenges dealing with an overwhelming amount of unstructured or semi-structured data in diverse formats, e.g., webpages represent petabytes of unstructured and semi-structured data [2]. In addition, documents and reports from specialized communities are constructed in their own native formats. Therefore, correlating and integrating diverse document collections has become a challenge and spurred research potentials in the area of knowledge discovery and data mining. For example, suppose a website stores blog posts as documents. A new blogger would like to search for a specific topic by formulating a query based on their information need. Based on the query the documents returned are the answer set. The blogger then sifts through all the documents in the answer set for their desired topic. Due to potential for enormous result sets and semantic ambiguities of words, this is an impractical solution and therefore a probable reason information retrieval (IR) techniques have become a popular research focus. This chapter reviews existing techniques regarding common information retrieval methods, relevance and retrieval evaluation, term categorization, ontology's, and various clustering techniques that when combined may provide these unique solutions.

2.1 Information Retrieval Models

The field of information retrieval aims to return documents in ranked order based on relevance of a document to a submitted query. The organization of the data in IR is usually unstructured, using natural language text and may be semantically ambiguous [4]. For example, the 20 Newsgroups data set is a collection of newsgroup documents containing unstructured natural language text and contains semantic ambiguities. To illustrate, suppose the term *plane* appears in the wood working newsgroup and *plane* also appears in the airplane newsgroup. The ambiguity is that others can refer to *plane* for wood working and others as to fly in a *plane*, requiring the pronoun to be resolved. Therefore, exploring the various IR models might be helpful in the retrieval of unstructured information. This section describes common IR models that have been used to help solve some IR issues, such as the Vector Space, Boolean, Extended Boolean, Probabilistic, Latent Semantic Indexing Analysis(LSI/A), probabilistic LSI (pLSI) and Latent Dirichlet Allocation(LDA) models.

2.1.1 Vector Space Model (VSM).   The Vector Space Model is an algebraic way of representing a document as vectors of term frequency counts. The documents are represented as vectors of term frequencies based on terms in the collection. Thus, each document contains terms which can be considered as dimensions in a multi-dimensional hyperplane which make mathematical comparisons much easier.

This is important so the similarity measures can be calculated. A query can also be represented as a vector of terms. Since the query is often short, the query vector will be extremely sparse. Given these two vector representations, we can measure their similarity using mathematical operations such as the cosine between the two (document and query).

Although straightforward in implementation, some adjustments are required to normalize terms and consequently weigh their importance in the document and the entire collection. A common method in VSM is to measure the frequency of terms or keywords k_i in document d_j from a corpus $\mathcal{D}$. The normalized term frequency is depicted as

$$tf(d,t) = \frac{freq_{t,d}}{max_\ell(freq_{\ell,d})} \tag{2.1}$$

where max_ℓ is the largest term frequency in d_i and the frequency of $d \in \mathcal{D}$ where $\mathcal{D}$ is the set of all documents in the corpus and $t \in \mathcal{T}$ where $\mathcal{T}$ is the set of all terms occurring in $\mathcal{D}$ [21]. A term that appears too frequently may be obsolete in terms of its relevance, so we determine the inverse term frequency to depict the importance of the term. For example, if a collection contains documents that are about *cats*, the animal, and the query term is *dozer*, the equipment, we want the documents that contain *dozer* to rank higher than those about *cats*. In this case, the term *cat* is obsolete since it will appear too frequently in the collection. So, we use the inverse term frequency *idf* defined as

$$idf = \log \frac{\mathcal{D}}{d_i} \tag{2.2}$$

where d_i is the number of documents that the term t_i appears [4].

Using VSM, the document vector can be defined as $\vec{d_j}=(w_{1,j}, w_{2,j},...w_{t,j})$. Therefore, for a given term i appearing in document j, the term weights $w_{i,j}$ are calculated by multiplying the term and inverse term frequencies to discount common terms.

$$w_{i,j} = \frac{freq_{t,d}}{max_\ell(freq_{\ell,d})} \times idf \tag{2.3}$$

Similarly, the term weights $w_{i,q}$ in the query q are weighted similar to the documents. Thus, the query vector can be defined as $\vec{q} = (w_{1,q}, w_{2,q},...w_{t,q})$, where each term of the vector can be calculated using [4]

$$w_{i,q} = 0.5 + \frac{freq_{t,q}}{(max_\ell)(freq_{\ell,q})}. \tag{2.4}$$

This representation creates *t*-dimensional vectors where their cosine angle can be treated as their similarity score and can be calculated as [4]

$$sim(d_j, q) = \frac{\vec{d_j} \cdot \vec{q}}{|\vec{d_j}| \times |\vec{q}|}, \tag{2.5}$$

$$= \frac{\sum_{i=1}^{t} w_{i,j} \times w_{i,q}}{\sqrt{\sum_{i=1}^{t} w_{i,j}^2} \times \sqrt{\sum_{i=1}^{t} w_{i,q}^2}}. \tag{2.6}$$

After the document and query vector representations are calculated, various IR models can be used to determine relevance ranking for queries over a given corpus. Using the vector space model, the cosine of the angle between the query vector and each document vector are calculated. This angle corresponds to how close the vectors are within the range of 0 to 1 where 0 means that the vectors are orthogonal, and 1 means they are "identical." The vector model has been compared to alternative ranking methods and the consensus was that the vector model is either superior or almost as good as the alternatives [4] by producing higher precision and recall values.

2.1.2 Boolean and Extended Boolean Model. The Boolean Model is designed to provide retrieval methods based on set theory and Boolean algebra [4]. The term weights are all binary where the term i appearing in document k, $\mathrm{w}_{ij} \in 0,1$, For instance, if a term from the query exists in the document the similarity of a document d_j and query q would be assigned 1, declaring the document as relevant and 0, otherwise. The query must express a Boolean expression which is not easy to translate from English into an information requirement. The Boolean model has greater precision in the area of data retrieval due to the binary decision, the data is there or it is not, based on relevance or non-relevance to the query. There are no criterion to determine a partial match based on the query. For example, for the query $q = k_i \wedge (k_j \vee k_l)$ and a document vector $\vec{d_j}=(0,1,1)$, the document will be considered non-relevant based on the query. This is because the query contains an *and* ($\wedge$) operator between the first and second term therefore the document's first and second term must both be 1's for it to be a relevant document. In this case the first term of the document is a 0 and the second term is a 1 and is deemed as non-relevant. The simplicity of this model, i.e., neglecting partial matching, leads to the retrieval of too few or too many documents.

The Extended Boolean attempts to refine the Boolean model by fractional weighting the terms and accounting for partial matches to retrieve a larger number of relevant documents [4]. Extended Boolean combines Boolean logic with VSM to improve retrieval performance and ranking over the Boolean model alone.

The similarity of a document d_j and query q are given by

$$sim(q_{or}, d_j) = \left(\frac{(x_1^p + x_2^p + ... + x_m^p)}{m} \right)^{1/p} \qquad (2.7)$$

7

and

$$sim(q_{and}, d_j) = 1 - \left(\frac{((1 - x_1^p) + (1 - x_2^p) + \ldots + (1 - x_m^p))}{m} \right)^{1/p} \qquad (2.8)$$

, where q_{or} is the *or* query and q_{and} is the *and* query. The p-norm model introduces p-distances where $1 \leq p \leq \infty$, by varying p between one and infinity the model changes the ranking from a vector ranking to a Boolean ranking.

A more generalized similarity formula can be applied recursively without regard to the number of AND/OR operators. For example, for a query $q = x_1 \ AND \ x_2 \ OR \ x_3$ the similarity between a document and query can be computed as

$$sim(q, d) = \left(\frac{\left(1 - \left(\frac{(1 - x_1)^p + (1 - x_2)^p}{2} \right)^{1/p} \right)^p + x_3^p}{2} \right)^{1/p}. \qquad (2.9)$$

The parameter p can have multiple values within the same query although the practical impact of this functionality are not known [4].

2.1.3 Probabilistic Model. The Probabilistic Model works from different set of assumptions, as only the user query and a set of documents deemed the relevant documents are compared [4]. These are referred to as the ideal answers, i.e., the query process entails specifying the properties, qualities in a document that relate to the query, of the ideal set [4]. The issue with this approach is that data properties of the corpus are not known, therefore, a guess or estimate is used to retrieve the first set of documents.

Once the results are given from the initial guess, the user reviews the retrieved documents, which can be a manual or automated process, and decides which documents are relevant and not relevant. This process is repeated numerous times until it is highly probable that the current set of documents becomes closer to the true desired document set.

The principle of the probabilistic model is that given a user query q and a document d_j, the model tries to estimate the probability that the user will find the document relevant. This probability is based on the query term and its relevant documents, which are a subset of documents from the collection that are relevant to the query term.

In the probabilistic model, the index term weights are binary where $w_{i,j} \in 0, 1$ and $w_{i,q} \in 0, 1$, depending on whether or not the term appeared in the document or query, respectively. The query q is a subset of index terms and R is the set of documents known to be relevant and $\overline{R}$ is the set of non-relevant documents. The probability that the index term k_i is present in the document randomly selected from R is $P(k_i \mid R)$ and the probability that it is not is given by $P(k_i \mid \overline{R})$. Therefore, measuring similarity is accomplished as [4]

$$sim(d_j, q) \approx \sum_{i=1}^{t} w_{i,q} \times w_{i,j} \times \left(\log \frac{P(k_i|R)}{1 - P(k_i|R)} + \log \frac{1 - P(k_i|\overline{R})}{P(k_i|\overline{R})} \right) \qquad (2.10)$$

.

Since $P(k_i|R)$ and $P(k_i|\overline{R})$ are initially unknown, they can be approximated where

$$P(k_i|R) = 0.5 \qquad (2.11)$$

and

$$P(k_i|\overline{R}) = \frac{n_i}{N} \qquad (2.12)$$

where N is the total number of documents and n_i is the number of documents in which k_i appear. Once the initial subset of relevant documents V and V_i are known, where V is the subset in which the term k_i appears, a baseline is estimated for the probabilities and new probability equations with an adjustment factor can be used [4]

$$P(k_i|R) = \frac{V_i + \frac{n_i}{N}}{V + 1}, \qquad (2.13)$$

9

$$P(k_i|\overline{R}) = \frac{n_i - V_i + \frac{n_i}{N}}{N - V + 1}.$$ (2.14)

The probabilistic model ranks the documents in decreasing order of probability of relevance of a document to the users need. The drawbacks of the model are the initial guess required for R and $\overline{R}$, the unaccounted term frequencies within documents, and the assumption that the index terms are independent.

2.1.4 Latent Semantic Indexing (LSI). Latent semantic indexing is a process similar to VSM which approximates a term-document matrix by one of lower rank using Singular Value Decomposition (SVD) [13]. The low-rank approximation of the matrix gives a new representation for the documents in the collection. The queries are cast into a low-rank representation which enables more efficient computation of the documents similarity score. Unlike the VSM, LSI addresses two major problems with VSM to include synonymies and polysemies. A synonymy refers to two words that have the same meaning such as *dog* and *canine* and a polysemy refers to a single word that has multiple meanings such as *bank*. Using the VSM to calculate the document and query vectors for a synonymy, for example *dog* and *canine*, the query vector q would contain *dog* and the document vector d would contain *dog* and *canine*. The problem lies in the calculation of the vector space which underestimates the true similarity of *dog* and *canine*. This is also true for polysemies where the VSM overestimates the similarity of *bank* [13]. Like VSM, LSI uses the cosine similarity to calculate the similarity of the document and query term but after SVD rank reduction which brings co-occurring terms closer together and thereby reducing the matrix dimensionality.

Singular value decomposition is a matrix decomposition method which produces matrices that are used in LSI with the end product being a low-rank approximation to the term-document matrix. SVD is the process of factoring a square matrix into the product of matrices which are derived from their eigenvectors. In SVD non-square matrix A is an $M \times N$ matrix where M is the number of terms in the corpus and N

10

is the number of documents in the corpus. Matrix A is then decomposed into three matrices U, S and V^T. The columns of U are the orthogonal eigenvectors of AA^T, S is the singular value matrix of A containing the principle components and the transpose of V, V^T whose columns are the orthogonal eigenvectors of $A^T A$ where A can be expressed by [13]:

$$A = USV^T \qquad (2.15)$$

In equation(2.15), U is a term by term matrix that depicts the relationships between the terms to include synonymies and polysemies where V is a document by document matrix that depicts the shared terms among the documents. S is a symmetric matrix containing the eigenvalues of U and V on its diagonal representing the term co-occurrences. Figure 2.1 depicts an example of the SVD matrices and their computed values with 3 documents and 3 terms where A is the term by document matrix, U is the term by term matrix, S is the co-occurrence matrix, and V is the document by document matrix.

Dimensionality reduction in LSI is done through a low-rank approximation of the SVD matrices where k is the value of the reduced rank. To truncate the full SVD matrices, the first k columns of U, the first k rows of V_T, and the first k rows and columns of S are kept, which are arranged in decreasing order. This truncation removes the noise by reducing dimensionality to expose the effect of the largest k singular values of the original SVD matrices. The reduced SVD in Figure 2.2 shows the matrices reduction where the shaded areas indicate the area of the matrix that is left after the k rows and columns are removed. After the reduction is complete, the new matrix A_k is computed by taking the product of U_k, S_k, and V_k^T which is the reduced rank approximation matrix.

The value of k should be chosen so as to minimize the Frobenius norm or Euclidean distance [34] which reduces the length of the vectors in matrix A. To approx-

11

Singular Value Decomposition			
A =	1.0000	2.0000	3.0000
	4.0000	5.0000	6.0000
	7.0000	8.0000	9.0000
U =	-0.4797	-0.7767	-0.4082
	-0.5724	-0.0757	0.8165
	-0.6651	0.6253	-0.4082
S =	16.8481	0.0000	0.0000
	0.0000	1.0684	0.0000
	0.0000	0.0000	0.0000
V =	-0.2148	0.8872	0.4082
	-0.5206	0.2496	-0.8165
	-0.8263	-0.3879	0.4082
U*S*V^T =	1.0000	2.0000	3.0000
	4.0000	5.0000	6.0000
	7.0000	8.0000	9.0000

Figure 2.1:   SVD example.

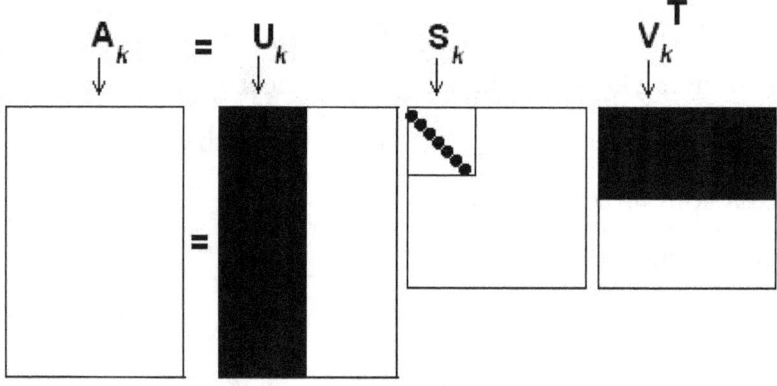

Figure 2.2:   Reduced SVD or Rank k Approximation

imate the best value for k with the smallest error use

$$A_k = \min \|A - A_k\|_F \qquad (2.16)$$

where A and A_k are both $M \times N$ matrices and F is the Frobenius norm. Therefore, using SVD for a given k, this approximation will yield the lowest possible Frobenius error.

The query will undergo SVD as well to produce a low-rank approximation that can be used in computing the document similarity scores. The original query vector $\vec{q}$ will be mapped to its LSI representation by reducing to k-dimensions and using

$$\vec{q}_k = \Sigma_k^{-1} U_k^T \vec{q}. \qquad (2.17)$$

Since the query $\vec{q}$ is just a vector of terms, new documents can be added to the collection by computing only $\vec{q}$ without recomputing the LSI representation. According to Garcia [13], the quality of the LSI matrices will degrade if too many documents are added since the co-occurrence of terms among documents will be ignored and recomputing the LSI representations is computationally expensive. Therefore, the original query vector $\vec{q}$ can be used in the cosine similarity measure since a query in the original space will be close to the documents in the k-dimensional space.

The computational cost of SVD is large, therefore LSI on a very large collection may not be feasible. Using a subset of a large collection and adding the remaining documents in is a work around but as the number of documents added increases, the quality of LSI decreases. In addition, LSI can be viewed as soft clustering due to the interpretation of each dimension in the reduced space as a cluster and the value of a document on that dimension as membership to that cluster.

2.1.5 Probabilistic Latent Semantic Indexing (PLSI). Probabilistic Latent Semantic Indexing (PLSI) is an automated indexing information retrieval model [20]. It is based on a statistical latent class model which is derived from LSI, making it a sounder probabilistic model. PLSI was introduced in 1999 by Jan Puzicha and Thomas Hofmann [20]. Unlike LSI, PLSI uses a statistical foundation that is more accurate in finding hidden semantic relationships [20]. The model uses factor analysis of count data, number of times an event occurs from a collection which is fitted from a training set of that collection. An Expectation Maximization (EM) algorithm solves the model to effectively find synonymy and polysemy relationships within a specific domain.

PLSI is based on the likelihood principle which is a principle of statistical inference which asserts that all of the information in a sample is contained in the likelihood function [20]. The statistical generative model called the *Aspect Model* is the basis of PLSI. The model is composed of the following probabilities

- select a document d with probability P(d),

- pick a latent class z with probability P($z|d$),

- generate a word w with probability P($w|z$).

The observed pair P(d, w) is the result of the generative model where the latent class z is discarded [20]. The observed pair is then given by a joint probability composed of

$$P(d, w) = P(d)P(w|d) \tag{2.18}$$

$$P(w|d) = \sum_{z \in Z} P(w|z)P(z|d). \tag{2.19}$$

The *Aspect Model* makes an independence assumption with the observed pair and conditional probabilities on the latent class z, where the words w are generated independently of the documents d. The latent class z is a variable that is used in predicting w conditioned on d where the word distributions are obtained by a convex combination of the aspects or factors P(w|z). The mixture of factor weights are characterized by the P($z|d$) which offers greater modeling power. Using Bayes' rule a

14

new version of the joint probability model is given by

$$P(w|d) = \sum_{z \in Z} P(z)P(w|z)P(d|z) \qquad (2.20)$$

which is just a reformatted version of the generative model.

To reduce word perplexity, model fitting must be accomplished through maximum likelihood estimation. The EM algorithm which involves two steps, the E-step and the M-step, is a standard procedure for maximum likelihood estimation [20]. The E-step computes the posterior probabilities of the latent variable z and the M-step updates the posterior probabilities computed in the E-step. The equation used in the E-step with a control parameter β is

$$P_\beta(z|d,w) = \frac{P(z)[P(d|z)P(w,z)]^\beta}{\sum_{z'} P(z')[P(d|z')P(w,z^{prime})]^{\beta'}_z}. \qquad (2.21)$$

The M-step equations are a convergent procedure that approaches a local maximum of the likelihood where the re-estimation equations are

$$P(w|z) = \frac{\sum_d n(d,w)P(z|d,w)}{\sum_{d,w'} n(d,w')P(z|d,w')} \qquad (2.22)$$

$$P(d|z) = \frac{\sum_w n(d,w)P(z|d,w)}{\sum_{d',w} n(d',w)P(z|d',w)} \qquad (2.23)$$

$$P(z) = \frac{1}{R} \sum_{d,w} n(d,w)P(z|d,w), R \equiv \sum_{d,w} n(d,w). \qquad (2.24)$$

PLSI uses query folding to incorporate queries into the *Aspect Model*. A representation of the query is computed in the EM iteration, where factors are fixed so that the mixing proportions P($z|q$) are adapted in each maximization step. The results are the probabilities and mixing proportions will have an affect on the term weights and the query will have a higher probability of matching the factors.

Query folding is the process of adding documents or queries that were not computed with the original training collection. This is done by fixing the $P(w|z)$ parameters and calculating the new query $P(z|q)$ by EM or Tempered EM. TEM is a model fitting algorithm that is closely related to deterministic annealing [20]. It is designed to solve the problem of over-fitting where noise overshadows the model relationships. If TEM is not used the model will perform less well on a folded-in query than on the data set used for training.

Similar to VSM and LSI, PLSI uses the cosine similarity metric to find the similarity between document vector representations to score the documents in the collection with regards to the query. The aspect vector for a query is generated by treating the query as a new document. The query is added to the model and the weights for the query are trained with the TEM algorithm. According to Hofmann, PLSI outperforms LSI with a precision increase of around 100% from the LSI baseline [20].

2.1.6 Latent Dirichlet Allocation (LDA). In 1990, de Finetti [7] established that any collection of exchangeable random variables has a representation as a mixture distribution, in general an infinite mixture. Thus, if we wish to consider exchangeable representations for documents and words, we need to consider mixture models that capture the exchangeability of both words and documents [7].

Latent Dirichlet Allocation (LDA) like PSLI is a generative probabilistic model for collection of discrete data such as a text corpora [7]. LDA is a three-level hierarchical Bayesian model where an item in a collection is modeled as a finite mixture over a set of latent topics. Topics are characterized by a distribution over the words in the corpus. The topics are then modeled as a finite mixture over a set of topic probabilities. The topic probabilities provide a reduced dimension representation of documents in a given collection.

The basic idea is that documents are represented as random mixtures over latent topics, where each topic is characterized by a distribution over words [7]. LDA uses the

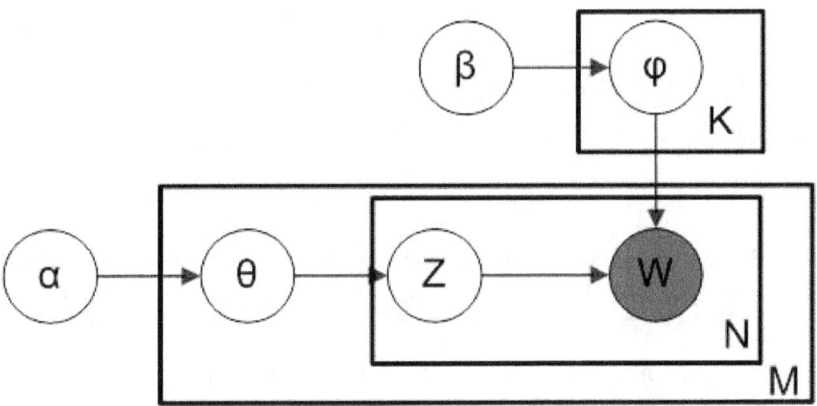

Figure 2.3:   Graphical model of LDA.

following process for each document d in the corpus D where we choose $N{\sim}$Poisson(ξ) and we choose $\theta{\sim}$Dir(α). For each of the N words w_n, $n=1...N$:

1. Choose topic z_n Multinomial(θ).

2. Choose a word w_n from P($w_n|z_n$, β), a multinomial probability conditioned on topic z_n.

There are several assumptions that are taken into account such as the dimensionality k of Dirichlet distribution is assumed to be known and fixed, the probabilities are parameterized by a $k \times V$ matrix β where β is estimated by $\beta_{ij}=$ P($w^i=1|z^j=1$). Also, N is independent of all other data variables θ and z. The k dimensional Dirichlet random variable θ takes values in the $(k$-1$)$-simplex and is given by the following density function [7]:

$$P(\theta|\alpha) = \frac{\Gamma(\sum_{i=1}^{k}\alpha_i)}{\prod_{i=1}^{k}\Gamma(\alpha_i)}\theta_1^{\alpha_1-1}...\theta_k^{\alpha_k-1} \tag{2.25}$$

where the parameter α is a k-vector with components $\alpha_i > 0$.

The probabilities of the entire corpus where the marginal probabilities of single documents are summed over the entire collection is given by:

$$P(D|\alpha,\beta) = \prod_{d=1}^{M} \int P(\theta_d|\alpha)(\prod_{n=1}^{N_d}\sum P(z_{dn}|\theta_d)P(w_{dn}|z_{dn},\beta))d\theta_d. \tag{2.26}$$

It has been shown that LDA outperforms other probabilistic models, such as the unigram, mixture of unigram and Probabilistic Latent Semantic Indexing (pLSI) models, for several document collections [7]. Latent Dirichlet Allocation is a generative probabilistic model for a collection of discrete data such as a text corpus. As shown in Figure 2.1.6, LDA is a three-level hierarchical Bayesian model where a document in a collection is modeled as a finite mixture over a set of latent topics K with a Dirichlet prior. Topics are characterized by a distribution over the words W in the corpus. The topics are then modeled as a finite mixture over sets of word-topic ϕ and document-topic θ probabilities. The topic probabilities provide a reduced dimension representation of the documents in the collection.

The plates/boxes represent repeated learning operations to obtain the various distributions. The variables α and β are parameters having a uniform Dirichlet prior representing the per-document topic distribution and the per-topic word distribution, respectively. Given a document i, θ_i represents its topic distribution. For each jth word in document i, z_{ij} represents its topic assignment. Note that w_{ij} is the only observable variable (shaded) and the rest are latent variables. Thus, inference of the various latent distributions is done using repeated Gibbs sampling—N times for each word in each document and M times for each document in the collection.

2.2 Document Preprocessing

Information retrieval methods require preprocessing of the documents to condition the date such as, eliminate non-essential data such as stopwords (common words), removal of suffixes and identifying index terms and/or keywords. It is sometimes important to remove unneeded punctuation, normalize numbers as well as date/time formats. All of these text preprocessing techniques are optional.

2.2.1 Lexical analysis.   Lexical analysis is the process of turning a stream of text into a stream of words. Spacing, punctuation and some abbreviations are deemed as non-essential in some IR arenas, therefore their removal may improve IR efficiency.

However, there are several instances when removal may deteriorate the context of the words. These cases are hyphens, certain digits (date and time), case of letters and punctuation. Therefore, their removal should be considered on a case-by-case basis.

Numbers are usually discarded as index terms, for example a query such as "boats that sank in 2005" with index terms *boats, sank, 2005* could retrieve more documents related to *2005* and not the *boats* that *sank*. In this case, the year *2005* would not be considered a good index term. However, when the numbers are interleaved such as *A.D. 200*, the number is important to the text. Also, numbers such as social security numbers, bank accounts or credit card numbers may be relevant and therefore should not be removed. The removal of numbers must be considered on a case by case basis and in the context of the collection. Finally, date and time should be normalized to the same format.

2.2.2 Stopword Removal. The focus of IR is to find the discriminatory words that will retrieve the most relevant documents. Stopwords are the frequent terms such as articles, prepositions and conjunctions are normally filtered as potential index terms. The removal of these terms minimizes the indexing structure and vector dimensions for a more streamline process.

The list of stopwords can be specialized to include verbs, adverbs and adjectives providing further index compression by eliminating uninteresting words. Although the goal may be to compress the index terms, this reduction can reduce the number of relevant documents retrieved. For example, if a user that is looking for documents containing the phrase "to be or not to be" and all the remains after stopword elimination is "be" it is impossible to properly recognize the documents that contain the specified phrase [4]. Most web search engines have opted for a full text index as to avoid ambiguities caused from stopword removal [4].

There are several common stopword lists that are available. The entire stopword list used in this thesis can be found in Appendix 1, however a comparison of two such lists are shown in Figure 2.1. Here the first column is a list containing 429 words

19

Table 2.1:   Comparison of Large and Small Stopword Lists.

List1	List2
a	a
-	a's
-	able
about	about
b	b
back	-
backed	-
backing	-
backs	-
be	be
...	...

and a second column is a list that is larger containing 571 words, the dashes indicate where the listings differ. Depending on the context in which the stopword list is used may dictate which list is appropriate, therefore editing the lists based on the domain is recommended.

2.2.3   Stemming.   Stemming is a procedure designed to reduce all words to their root by stripping each word of its derivational and inflecational suffixes [32]. This process is useful when counting word frequency, matching words with suffixes is often less successful than finding matching stemmed words. In the areas of computational, information retrieval, and mathematical analysis, word stemming is essential in the evaluation of terms and keywords [32].

Several algorithms are used to perform stemming, each with its own benefits. Stemming algorithms may have semantical implications and therefore should be used with caution. Sometimes suffixes provide clues to the grammatical context of words and should not be removed. For example, the word *cardiology* could be stemmed to *cardio* which could have several possible suffixes such as *cardiology, cardioprotectant, cardiopulmonary/ cardiovascular.* Stemming these words to the root *cardio* may not be a good idea since there are many word forms in very different domains. Therefore,

20

selecting the appropriate stemming algorithm for the collection is important, but most IR algorithms do not use stemming for these reasons [32].

According to Yates [4], there are four types of stemming algorithm; affix removal, table lookup, successor variety and n-grams. Table lookup stores a table of all index terms and their stems, so terms from queries and indexes could be stemmed very fast. The successor variety determines word and morpheme boundaries and using one of the following methods cutoff, peek/plateau or complete method to find the stem word. N-grams uses the identification of diagrams and trigrams as its basis and then association measures are calculated between pairs of terms based on shared unique diagrams. N-grams stemming is more of a clustering algorithm using matrix to store similar words and then uses a single link clustering method.

Affix removal stemming is the simplest and can be implemented efficiently, Porter's algorithm is the most common affix removal stemming algorithm. The Porter algorithm uses a suffix list and applies a series of rules to the suffixes of the words in the text [4]. An example of one of the rules is

$$s \longrightarrow \phi \tag{2.27}$$

which converts plural forms to their singular forms by substituting s by nil, ϕ. Furthermore, applying the following rules

$$sses \longrightarrow ss \tag{2.28}$$

$$s \longrightarrow \phi \tag{2.29}$$

to the word possesses yields the stem word possess.

2.2.4 Identify index terms or keywords. Index terms or keywords are the unique words that remain after the text pre-processing is complete. The terms remaining are usually nouns due to the elimination of verbs, adverbs, adjectives, connectives

21

articles and pronouns during pre-processing. During the parsing process several nouns that appear near each other can be grouped into a topic. The topics formed are called noun groups. The distance between the nouns is a predefined measure, usually the number of words between the nouns. These noun groups can be used as the index terms.

In some cases full-indexing is used which incorporates the entire vocabulary. For specialized domains the index terms may be pared down by a subject matter expert to narrow the index term scope. The process of selecting index terms can be accomplished manually for specialized areas but automating term selection is a common practice.

The vocabulary ultimately defines a thesaurus of index terms. The thesaurus consists of a pre-compiled list of important words in a given domain of knowledge and for each word in this list a set of related words [4]. The purpose of the thesaurus is to provide a standard vocabulary for indexing and searching and to assist users with identifying query terms and for query reformulation. A thesaurus may be used in query reformulation. A user determines the information that they requires and an IR system can provide a thesaurus to narrow the search terms based on conceptualizing the query. Since a user may not select the correct terms for searching based on a lack of experience, the IR system can assist the user by providing related terms based on the query. On the other hand, a thesaurus can be used with the initial query but this requires expensive processing time since the thesaurus has not been tailored to a query. Therefore, a thesaurus may not be computationally efficient especially in IR systems where the user expects fast processing.

2.3 Query Operations

A query operation is a precise request for information from a collection. The query can be composed of free text such as web search engines or in a computer language for databases or other information systems.

Initial query formulation is usually done with little knowledge of the collection therefore it may be difficult for users to describe a well-formed query for effective retrieval. This idea implies that users spend the majority of their retrieval time reformulating queries. Furthermore, relevance feedback can assist users in query reformulation, query expansion and reweighting query terms.

2.3.1 Query Expansion and Term Reweighting. Ad-hoc retrieval relies on the user's query to provide a variety of terms and varying term frequencies to differentiate term importance [27]. There are two stages in the ad-hoc retrieval process, the first is the initial user query and the second is the expansion of the query based on the relevant documents retrieved using the initial query.

The initial retrieval of n best-ranked documents are regarded as relevant without user interaction. They are then used to train the initial query term weights and expand the query. The expanded query is used in a second retrieval attempt can give better results than the initial query if the initial results are reasonable and has some relevant documents within the best n. The process will work properly only if the initial query contains a variety of terms and term importance is evident.

This is where term weighting can improve retrieval results by calculating a modified query $\vec{q}_m$. The three classic methods to calculate the modified query $\vec{q}_m$ based on relevance feedback are

$$Standard\_Rochio : \vec{q}_m = \alpha\vec{q} + \frac{\beta}{|D_r|} \sum_{\forall \vec{d}_j \in D_r} \vec{d}_j - \frac{\gamma}{|D_n|} \sum_{\forall \vec{d}_j \in D_r} \vec{d}_j \qquad (2.30)$$

$$Ide\_Regular : \vec{q}_m = \alpha\vec{q} + \beta \sum_{\forall \vec{d}_j \in D_r} \vec{d}_j - (\gamma)\Sigma_{\forall \vec{d}_j \in D_n}\vec{d}_j \qquad (2.31)$$

$$Ide\_Dec\_Hi : \vec{q}_m = \alpha\vec{q} + \beta \sum_{\forall \vec{d}_j \in D_r} \vec{d}_j - (\gamma)max_{non-relevant}(\vec{d}_j) \qquad (2.32)$$

where $max_{non-relevant}(\vec{d}_j)$ is a reference to the highest ranked non-relevant document. The documents in D_r and D_n are those that the user deemed as relevant or non-relevant. The current understanding is that the three techniques yield similar results, however in the past, Equation (2.32) was considered slightly better [4].

2.3.2 Query Expansion Through Local Clustering. In IR, clustering is the practice of grouping common documents into subsets so further analysis can be accomplished on their relationships [4]. Clustering is another common way to expand queries. This technique uses association matrices to quantify term correlations such as term co-occurrence and to use those terms to expand the query. The problem with the association matrices is that they do not adapt well to the current query. Several local clustering techniques may be used to alleviate this issue by optimizing the current search. There are three techniques discussed to include association, metric and scalar.

Association clustering [4] uses an association matrix composed of co-occurring terms within the documents. The association comes from the notion that co-occurring terms inside documents tend to have synonymity association. Therefore, a matrix $\vec{s}$ is developed using the terms as the s_i rows and the documents d_j as columns where the matrix values represent the co-occurring frequency of the terms. Then they are clustered by taking the u-th row of the matrix $\vec{s}$ and returns the set $S_u(n)$ of n largest values of $s_{u,v}$, where u and v ($u{\neq}v$) are the values in the matrix, v varies over the set of local terms. $S_u(n)$ is then a local association cluster around the term s_u [4].

Second is the metric clustering technique [4] which takes into account where the terms occur in the documents not just their co-occurrences. The distance $s(k_i, k_j)$ between terms k_i and k_j are the number of terms that are between them. Metric clustering is similar to association clustering but uses distance $s(k_i, k_j)$ for the values in the matrix. They are clustered by taking the u-th row of the matrix $\vec{s}$ and returns the set $S_u(n)$ of n largest values of $s_{u,v}$, where u and v ($u{\neq}v$) are the values in the matrix, v varies over the set of local terms. $S_u(n)$ is then a metric correlation cluster around the term s_u [4].

24

Lastly, scalar clustering [4] is similar to association clustering where it finds synonymy relationships between terms using term neighborhoods. This is done by using term neighbors, where terms with similar neighbors are most likely to have a synonymy relationship. To quantify the relationship the terms are split into two vectors where $\vec{s}_u$ and $\vec{s}_v$ ($u \neq v$) are correlated terms values. The cosine angle between the vectors is used to induce a similarity value. Like the last two clustering methods the set $S_u(n)$ of n largest values of $s_{u,v}$ where u and v ($u \neq v$) are the values in the vectors and is a scalar cluster around s_u [4].

The clusters produced from these techniques are the terms that are used in the expanded query $\vec{q}_m$. There are a couple of ways to do this which are adding the terms in the clusters to the original query or replacing the original query with the clustered terms. Either option will provide an expanded query to retrieve relevant documents based on these clustering techniques.

2.3.3 Relevance Feedback. Relevance feedback [4] is a very important tool, especially if a collection is unlabeled. The initial results a query returns may not reflect the desired output this is where user feedback comes in handy. This can be done using two relevance feedback methods which are global and local. Global methods expand the user query or reformulates the query terms based on the initial result set. These methods may include the incorporation of a thesaurus, the generation of a thesaurus, or spelling correction [34].

Local methods adjust the query based on initial returned documents. The local methods include relevance feedback, pseudorelevance feedback and indirect relevance feedback. According to Yates, relevance feedback (RF) is the most used and most successful approach to improve IR results [34].

Relevance feedback involves the user to improve the relevancy of the returned documents in the initial results set. The RF process proceeds as follows:

1. User issues query.

2. System returns an initial set of results.

3. User marks returned documents as relevant or nonrelevant.

4. System retrieves a better set of results based on the user input.

5. System displays a new set of relevant documents.

This process can be repeated several times where relevancy may be improved with each iteration.

A common algorithm for implementing RF is the Rocchio algorithm [34]. Rocchio uses the vector model and combines it with the relevance feedback information provided by the user. The goal of the algorithm is to maximize the similarity of relevant documents while minimizing the similarity of non-relevant documents. The optimal query vector is the equation on which the Rocchio algorithm is based:

$$\vec{q}_{opt} = \arg\max_{\vec{q}}[sim(\vec{q}, C_r - sim(\vec{q}, C_{nr})] \qquad (2.33)$$

where $\vec{q}$ is the query vector and C_r are the relevant documents and C_{nr} are the non-relevant documents.

The problem with the optimal query vector is that the full set of relevant documents is not known. This is where Rocchio's algorithm modifies the query vector $\vec{q}_m$ with weights attached to each term. and D_r the relevant documents and D_{nr} are the non-relevant documents. Therefore, $\vec{q}_m$ is given by:

$$\vec{q}_m = \alpha\vec{q}_0 + \beta\frac{1}{|D_r|}\sum_{\vec{d}_j \in D_r}\vec{d}_j - \gamma\frac{1}{|D_{nr}|}\sum_{\vec{d}_j \in D_{nr}}\vec{d}_j. \qquad (2.34)$$

2.4 Relevance Retrieval Evaluation

The final step in any information retrieval process is to evaluate the results and determine their usefulness. There are several measures to determine if the results are useful such as user feedback, precision and recall. This section discusses methods for improving relevancy results through user feedback as well as the measures of relevancy, namely precision and recall.

2.4.1 Precision and Recall. The goal in any IR system is to maximize both precision and recall. Precision is the percentage of retrieved documents which are relevant and recall is the percentage of relevant documents retrieved. Let R be a set of relevant documents and A be an answer set to a retrieval request I and Ra be the number of documents in the intersection of R and A. Precision is then calculated by:

$$Precision = \frac{|Ra|}{A} \tag{2.35}$$

and recall is calculated by:

$$Recall = \frac{|Ra|}{R}. \tag{2.36}$$

Figure 2.4 is an example of a recall and precision graph from the Trec08 conference [45]. The graph shows the concave shape where the area under the curves represent the average precision. Moving the curves up and out to the right depicts the improvement of both precision and recall, making the average precision increase. These measures assume that the relevant documents are known, which may not be the case. Therefore, variations of these equations such as F-measure and R-precision are used to determine the true curve of precision versus recall. This comparison is created by averaging the results over various queries. However, this does not paint a clear picture of the results of individual queries or the algorithms used in the IR system.

A common single value measure is called the R-Precision method where a single value summary of the ranking is used. This is done by computing the precision at the R-th position in the document ranking where R is the number of relevant documents int the collection for a given query. R-precision can also be averaged over the entire set of queries.

R-Precision is a useful single value measure to compare various IR algorithms and determine which method outperforms the rest. Creating an average recall ver-

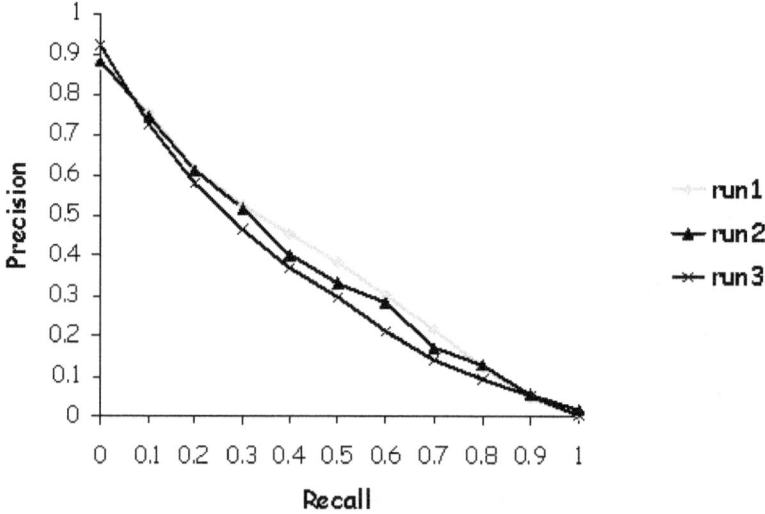

Figure 2.4:   Typical Precision and Recall graph

sus precision evaluation strategy for IR systems are used extensively in IR retrieval literature [4].

2.4.2 Perplexity.   Perplexity, a common metric for evaluating natural language processing models [25], is used to evaluate the models. The perplexity value computed on the held-out test data indicates how well the model is able to generalize the unseen data. The lower the perplexity the better the model is able to generalize. The following equation is used for computing perplexity:

$$perplexity(D_{test}) = exp\left\{-\frac{\Sigma_{d=1}^{M}log\,p(\mathbf{w}_d)}{\Sigma_{d=1}^{M}N_d}\right\}\qquad(2.37)$$

where D_{test} is the set of test documents held out from the collection, M is the number of documents in the collection, $p(\mathbf{w}_d)$ is the probability of the words in document d, and N_d are the number of words in document d. Notice that the numerator, is the entropy of the collection, given $p(\mathbf{w}_d)$, for each document [7].

28

2.5 Text Categorization and Ontologies

An ontology is a formal representation of a set of concepts within a domain and the relationships between those concepts. Ontologies are used in the information retrieval domain to model lexical and domain knowledge and for information extraction. The common ontologies used in natural language processing are WordNet and the Web Ontology Language (OWL).

2.5.1 WordNet. WordNet is a lexical database that links English nouns, verbs, adjectives, and adverbs to sets of synonyms called synsets which are then linked by their semantic relations (Antonymy, Hyponymy, Meronymy, and Troponymy) which determine word definitions [37]. Where the Web Ontology Language's aim is information organization, WordNet is used to provide semantic relationships by linking words to their semantic counterparts, improving relevancy ranking [21].

WordNet uses language definitions as a set W of pairs (f,s), where f is a word string and s is the set of meanings for that string where f can be used to express a particular s [37]. Currently, there are 118,000 word forms, a string over a finite alphabet, and more than 90,000 word senses, an element from a given meaning set, with more than 166,000 (f,s) pairs [37]. The semantic relationships between word forms or synsets are connected by pointers between word forms or (f,s) pairs. This provides a link between a word and the various synonyms, antonyms, and various other semantic relationships useful in defining a broader scope of semantic relations. The additional relationship information provided by WordNet aids in the semantic contexts that other IR methods fail to consider such as the troponomy march or walk, the manner in which one moves. Incorporating WordNet with common IR methods such as vector modeling or document clustering could be beneficial to the IR community [37].

WordNet has been used in many areas as a reference ontology in the area of information retrieval. For instance, Varelas et al. uses WordNet to detect similarities that are semantically but not lexicographically related. They combined their approach with a novel IR method resulting in a better performance over other IR methods [44]. Hearst devised a method for automatic discovery of WordNet relations by searching for lexico-semantic relations shown to be useful in the detection of hidden semantic relationships [18]. Several others have used WordNet for text classification to include Rosso et al. [24], Chua and Kulathuramaiyer [11], and Mansuy and Hilderman [35]. Measuring concept relatedness is another area where WordNet has made contributions, eg., automatically annotating text with cohesive concept ties [42] and measuring relatedness of pairs of concepts [43]. WordNet has also been paired with Roget's and Corpus-based Thesauri to augment WordNet's missing concepts [33]. Text clustering algorithms have been enhanced using WordNet as shown by Liu et al. [30] and Hotho et al. [21]. These works have shown that ontology augmentation is useful in identifying hidden semantic relationships and is worth investigating for improving LDA results.

2.5.2 Web Ontology Language (OWL). Information contained on the World Wide Web, a corpus where there is less definable structure requires some concrete organization to attain useful knowledge. One way that this can be done is to form computational patterns that can be connected in such a way that meaningful information can be extracted.

The Web Ontology Language (OWL) enables the mining of the massive amount of unorganized data that would otherwise be meaningless [3]. OWL is a language that makes use of ontologies. OWL is based on defining and instantiating Web specific ontologies. Ontology is defined as "That department of the science of metaphysics which investigates and explains the nature and essential properties and relations of all beings, as such, or the principles and causes of being" [1]. In computer science and information science ontology is a formal representation of a set of concepts within a domain and the relationships between those concepts. In OWL, ontology's define classes, properties and their instances. Given such an ontology, the OWL formal semantics specifies how to derive its logical consequences, or entailed from its semantics. These entailments may be based on a single document or multiple distributed documents that have been combined using defined OWL mechanisms.

OWL differs from other Web standard languages as it makes use of the semantics to create a useful tool outside of its' original function. For example, the Extensible Markup Language (XML) is in a message format rather than a knowledge representation that OWL provides. Furthermore, OWL consists of three sub languages; OWL Lite, OWL DL, and OWL Full. Each of these sub languages designed to provide specific user requirements.

OWL Lite is designed for lite user's, as the title expresses. Users that require classification hierarchy and simple constraint features where cardinality can be expressed as 0 or 1. OWL Lite also provides swift migration from other taxonomies.

OWL Description Logic (DL) is for the user that requires the maximum computational completeness and are guaranteed to return a computation. DL provides all the functions of the OWL language restricting type separation(classes and properties can not be one in the same). OWL DL is designed to support the existing DL business segment and has desirable computational properties for reasoning systems.

OWL Full, is designed for the user who is not concerned with computational guarantees but are interested in strict expressiveness. Also, OWL Full allows an ontology to augment the meaning of a predefined vocabulary.

The OWL structure is based on the formal syntax and semantics which are an extension of the Resource Description Framework (RDF) [3]. RDF is an assertion language on which OWL is based. It provides a means to express propositions using precise formal vocabularies. OWL uses RDF to specify the specific vocabularies to be used. An XML file with an RFD tag provides the necessary identifiers to provide a meaningful and readable ontology. Figure 2.5 depicts the use of OWL to add comments, version control, importing existing ontologies and labeling [3].

Since ontologies are like software they change over time which require version control and OWL provides a version definition function to link versions together and

```
<owl:Ontology rdf:about=" " />
 <rdfs:comment>An example OWL ontology</rdfs:comment>
  <owl:priorVersion rdf:resource="http://www.w3.org/TR/2003/PR-owl-guide-20031215/wine" />
  <owl:imports rdf:resource="http://www.w3.org/TR/2004/REC-owl-guide-20040210/food" />
 <rdfs:label>Wine Ontology</rdfs:label>
```

Figure 2.5:   Example of Owl Structure

track history of an ontology. This along with other functions that OWL provides enable users to create ontologies that are easily linked and make searching the Web that much easier [3].

The data described by an OWL ontology is interpreted as a set of "individuals" and a set of "property assertions" which relate these individuals to each other. An OWL ontology consists of a set of data types which place constraints on sets of individuals that make up classes and the types of relationships permitted among two objects. These data types provide semantics by allowing systems to infer additional information based on the data explicitly provided.

OWL is a useful tool in the area of information retrieval, by using the relationship data provided by OWL ontologies. The use of OWL strives to create organization that provides a means to obtain useful information from complex relationships which would otherwise be overlooked. The use of software agents with an unorganized ontology provides suboptimal results. Using OWL or another ontology language to organize data and creating domain specific ontologies which may improve information retrieval results since information will have a defined structure.

2.6   Document Clustering and Visualization

The document clustering hypothesis states that documents within the same cluster behave similarly with respect to relevance to some information needs [34]. Therefore, a cluster with a document that is relevant to the search criteria may also contain other documents that are relevant, which is the sole purpose of document clustering, gathering documents with similar terms. Document clustering is a form of unsupervised learning, i.e, where no human expert has labeled or assigned the documents to classes [34]. In this way, the learned algorithm and corpus will determine which clusters the documents appear. The clustering algorithms used to create clusters use a distance measure often a Euclidean distance, which is the distance of documents from their cluster centers. There are two common clustering algorithms, k-means and hierarchical clustering, which are briefly discussed in the following subsections.

2.6.1 K-means Clustering. K-means clustering is a flat clustering algorithm whose objective is to minimize the average squared Euclidean distance from a given cluster ω having centroid $\vec{\mu}$ where $\vec{x}$ is the length normalized documents [34]. The centroid is given by:

$$\vec{\mu}(\omega) = \frac{1}{|\omega|} \sum_{\vec{x} \in \omega} \vec{x}. \tag{2.38}$$

Each cluster in K-means should be a sphere with the centroid at the center of gravity and the clusters should not overlap [34]. To measure the effectiveness of the K-means clustering the residual sum of squares(RSS) is calculated. RSS is the squared distance from each centroid summed over all of the vectors, which is formulated as follows:

$$RSS_k = \sum_{\vec{x} \in \omega} |\vec{x} - \vec{\mu}(\omega_k)|^2. \tag{2.39}$$

The objective function of K-means is RSS and minimizing it is equivalent to minimizing the average squared distance. This gives a measure of how well the centroids represent their documents [34].

2.6.2 Hierarchical Clustering. Hierarchical clustering provides a structured output which is more informative than other clustering algorithms [34]. As opposed to K-means the hierarchical clustering algorithms do not require specifying the number of clusters ahead of time. Although hierarchical algorithm are deterministic, their efficiency is less desirable having a complexity of at least quadratic compared to K-

means which is linear. Therefore, these trade offs must be considered when choosing a clustering algorithm.

Hierarchical clustering algorithms are either top-down or bottom-up. Bottom-up treats each document as a singleton and merges pairs of clusters until all documents are in a single cluster. Top-down splits the clusters until individual documents are reached. Bottom-up is more frequently used in IR [34].

There are four common hierarchical clustering algorithms: single-linkage, complete-linkage, group-average, and centroid. Single-linkage is the similarity of two clusters by their most similar members and where the two clusters are closest together which is a local criteria. The clusters are merged based on the two closest pairs and then by the next closest pair. A single-linkage clustering side-effect called chaining occurs, where documents are added to the cluster and can create a chain effect. This chaining effect can produce a straggling cluster which can be extended for long distances, an undesirable side-effect.

Complete-linkage based clustering solves the issues that single-linkage based clustering creates but produce other structure irregularities caused by outliers. Complete-link based clustering is the similarity of two clusters by their most dissimilar members. The merge criteria are non-local and take into account the cluster structure and thereby reduce the chain-effect that is produced by single-linkage clustering.

Group-average clustering is another approach that takes into account all similarities between documents which elevates the problems that arise with single and complete-linkage clustering algorithms. The group-average clustering algorithm averages the similarity of all the pairs of documents to include those in the same cluster but self-similarities are not included in the average.

The final hierarchical clustering algorithm is the centroid clustering algorithm. The similarity of two clusters is based on the similarity of their centroids [34]. This is similar to the group-average algorithm except that centroid clustering excludes pairs that are in the same cluster. Centroid clustering is more commonly used because it is simpler to calculate the similarity of two centroids than to calculate the pairwise similarity in group-average clustering.

2.6.3 Self-Organizing Maps (SOM). Self-organizing maps is a type of artificial neural network that is trained using unsupervised learning to produce a low-dimensional (typically two-dimensional), discretized representation of the input space of the training samples, called a map. SOM's consist of a fixed lattice where multi-dimensional data is represented in a 2D space. Self-organizing maps are different than other artificial neural networks in the sense that they use a neighborhood function to preserve the topological properties of the input space [26].

A self-organizing map consists of components called nodes or neurons known as processing elements. Associated with each node is a weight vector of the same dimension as the input data vectors and a position in the map space. The usual arrangement of nodes is a regular spacing in a hexagonal or rectangular grid. The procedure for placing a vector from data space onto the map is to find the best-matching unit in a vector to the vector taken from the data space and to assign the map coordinates of this node to the vector. The best-matching unit can be found using

$$c = argmin_i||x - m_i|| \tag{2.40}$$

where c is the index of the best-matching unit, x is an input from the input sample, and m_i is the vector associated with the processing element i, and $||.||$ is the distance metric.

After c has been calculated, c and all of the m_i's with a certain geometric distance in the map space (physically the grid) can be updated using

$$m_i(t + 1) = m_i(t) + \alpha(t)h_{ci}(t)[x(t) - m_t] \tag{2.41}$$

where $t \geq 0$ is a discrete coordinate, $\alpha(t)$ is a monotonically decreasing learning rate and h_{ci} is a neighborhood function. In order for convergence to occur h_{ci} must

Figure 2.6:   SOM Cluster Map. [36]

approach zero with increasing time and acts as a smoothing kernal over the SOM lattice to ensure the converged map is ordered [26].

Self-organizing maps naturally cluster so that the data with similar features are mapped to the same or nearby processing elements [36]. The topology of the input space is preserved on the lattice, i.e, relationships between samples in the high-dimensional input space are preserved on a low-dimensional mapping [26]. This preservation makes the SOM a great visualization tool to map multi-dimensional data to a 2D representation, as seen in Figure 2.6.

III. Augmenting LDA and Rank Threshold Detection using WordNet

In an ever-increasing data rich environment, actionable information must be extracted, filtered, and correlated from massive amounts of disparate often free text sources. The usefulness of the retrieved information depends on how we accomplish these steps and present the most relevant information to the analyst.

It has been shown by Blei et al. that LDA outperforms other probabilistic models such as the unigram, mixture of unigram and Probabilistic Latent Semantic Indexing models, for several document collections [7]. Latent Dirichlet Allocation is a generative probabilistic model for collection of discrete data such as a text corpus. LDA is a three-level hierarchical Bayesian model where an item in a collection is modeled as a finite mixture over a set of latent topics. Topics are characterized by a distribution over the words in the corpus. The topics are then modeled as a finite mixture over a set of topic probabilities. The topic probabilities provide a reduced dimension representation of documents in a given collection.

Figure 3.1 depicts the general document modeling process, where the collection is encoded to include text processing, use of ontologies and query introduction. After encoding, the modeling process can be accomplished with various modeling algorithms to include LDA, SOM, LSI, PLSI, Vector or Boolean. Finally, the results are presented to depict the output of the modeling algorithm. This model is tailored later for LDAWN in Section 3.2.

3.1 Process Overview

The principle advantages of generative models, such as LDA, include their modularity and their extensibility. They are easier to modify and study; for example, using an alternative sampling method from Gibbs Sampling as used in finding scientific topics [15] to Random Sampling used in face recognition [23]. It is also possible for LDA to be embedded in complex models as well as extending LDA by introducing background knowledge to improve word and topic distributions, as we do here.

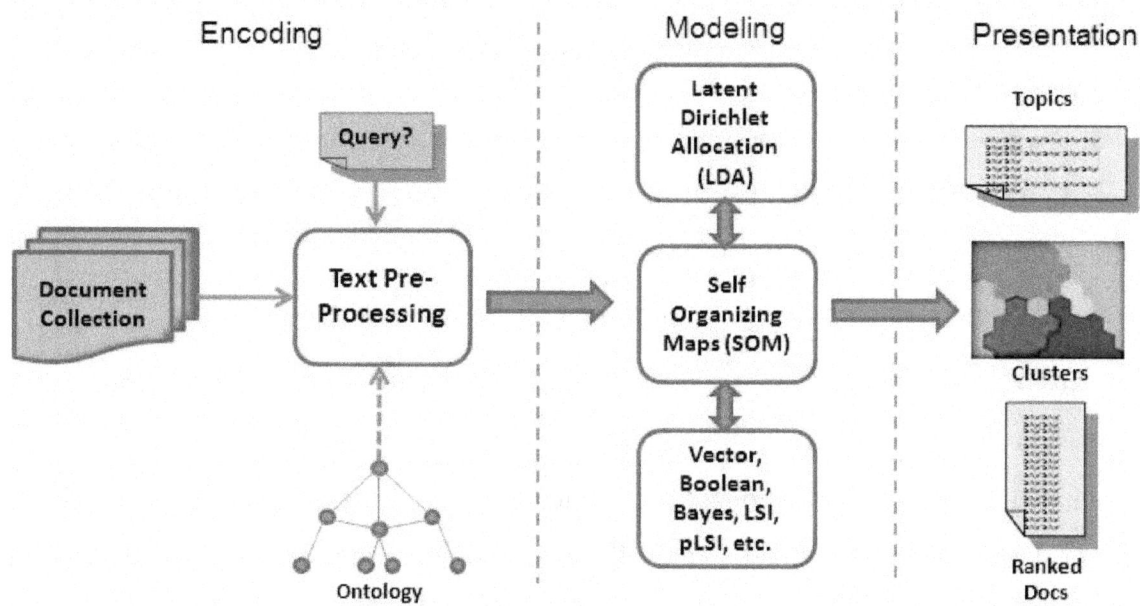

Figure 3.1:   General document modeling process.

Recent work in this area include clustering and visualization using LDA and Self-Organizing Maps (SOM) [36], document modeling using probabilistic topic models [7], and a comparison of probabilistic topic models [41]. Both Blei's et al. and Styvers' and Griffiths' work aim to analyze the contents of documents and the meaning of words using probabilistic topic models. Their results show that LDA outperforms other probabilistic models. Additionally, Millar et al. [36] work shows how LDA and SOM's can be used together to cluster and visualize topic distributions, their results on the 20 Newsgroups and NIPS collections showed good behavior. However, they pointed out a couple of challenges such as setting LDA hyperparameters and choosing a reasonable topic number. Our approach to LDA will not differ, but the novelty of our approach comes with the incorporation of WordNet into the LDA document modeling process.

WordNet has been used in many areas as a reference ontology in the area of information retrieval. For instance, Varelas et al. [44] uses WordNet to detect similarities that are semantically but not lexicographically related. They combined their approach with a novel IR method resulting in a better performance over other IR methods [44]. Hearst devised a method for automatic discovery of WordNet relations by searching for lexico-semantic relations shown to be useful in the detection of hidden semantic relationships [18]. Several others have used WordNet for text classification to include Rosso et al. [24], Chua and Kulathuramaiyer [11], and Mansuy and Hilderman [35]. Measuring *"concept relatedness"* is another area where WordNet has made contributions, i.e., automatically annotating text with cohesive concept ties [42] and measuring relatedness of pairs of concepts [43]. WordNet has also been paired with Roget's and Corpus-based Thesauri to augment WordNet's missing concepts [33]. Text clustering algorithms have been enhanced using WordNet as shown by Liu et al. [30] and Hotho et al. [21]. These works have shown that ontology augmentation is useful in identifying hidden semantic relationships and is worth investigating.

3.2 *Augment LDA using WordNet (LDAWN)*

Tools used in document modeling face difficult challenges dealing with data management and data diversity compounded with the overwhelming amount of unstructured or semi-structured data. As an example, web pages represent petabytes [2] of unmanageable amount of semi-structured data. In addition, various documents and reports from specialized communities are constructed in these formats. As a result, many research activities are flourishing in the area of knowledge discovery and data mining of various document collections.

3.2.1 *LDAWN Problem Definition.*   In the area of Knowledge Discovery and Data Mining (KDD), data management systems often become overwhelmed with source documents, in free text form, that are not labeled or pre-assigned to specific topics. The usefulness of the retrieved information depends on how we accomplish these steps and present the most relevant information to the analyst.

3.2.2 *LDAWN Goals and Hypothesis.*   One method for extracting information from free text is Latent Dirichlet Allocation (LDA), a document categorization technique to classify documents into cohesive topics. Although LDA accounts for some

implicit relationships such as synonymy (same meaning) it often ignores other semantic relationships such as polysemy (different meanings), hyponym (subordinate), and meronym (part of). To compensate for this deficiency, we incorporate explicit word ontologies, such as WordNet, into the LDA algorithm to account for various semantic relationships.

The benefit of supplementing the LDA algorithm with WordNet synsets is to introduce semantic relationships LDA is not designed to discover such as polysemes, hyponyms, meronyms, troponomys etc. For example, a document about *dog* may not be related to a document about *cat* by the LDA algorithm but their semantic ties with *animal* can reveal their hidden relationships. To avoid further complicating the LDA algorithm as it reduces the term-document matrix to a much smaller word-topic categorization, any enhancement should not increase the dimensionality of the problem space. LDAWN, achieves both of these objectives.

The LDAWN algorithm increases document term frequencies by incrementing terms by the number of new entries for WordNet terms appearing in the same document. This is a similar strategy to the "add strategy" used by Hotho et al. [21], where a term that appears in WordNet as a synset is accounted for at least twice but could be accounted for more often due to terms having more than one synset. As a result, term frequencies are increased for words contained in a document that have semantic relationships with other words contained in the same document. This in turn increases the LDA word-topic distribution probabilities for a given word with semantic relations that have affected its term frequency count. The term frequency directly affects the LDA probability for $p(w|z)$, where each topic z is characterized by a distribution over the words w. This alters the distribution so that some words are more probable than others, therefore identifying words that are related and that better fit the word-topic distribution. For example, if the word *dog* appears twice in a document and the term *canine* appears once in the same document, the term frequencies for those terms are incremented by the number of occurrences. Therefore, the term frequency for *dog* and *canine* are both three for that document. This method

40

Table 3.1:   LDAWN and LDA word-topic distributions for *dog* and *canine*.

LDA Topic 53:	canine	allergens	supply	seen	responds	gmt	relative	dogs
LDAWN Topic 96:	dogs	canine	allergens	supply	seen	responds	gmt	relative

gives equal word probability for both *dog* and *canine* thereby explicitly defining their semantic relationship.

This method directly affects LDA in terms of computing the posterior distribution of hidden variables, which is intractable. Therefore, using variational inferences to formulate the computation of a marginal or conditional probability, a family of distributions on the latent variables are obtained, making the computation tractable. One of these distributions is the variational distribution which is a conditional distribution, varying as a function of $\mathbf{w}$, where $\mathbf{w}$ are the words in the distribution. Since the variational distribution is explicitly dependent on $\mathbf{w}$, increasing the probability's in $\mathbf{w}$ directly affects the variational distribution which in turn influences the word-topic distribution. To show how the word-topic distribution is affected by LDAWN a comparison of the LDAWN and LDA distributions show the affects of incorporating an ontology. The LDAWN distribution should have higher probabilities for semantically related terms than LDA for the topics they are assigned. Using a small collection of documents pertaining to *dogs*, Table 3.1 are the LDAWN and LDA word-topic distributions for the words *dog* and *canine* in order of word probability, from greatest to least.

The LDAWN word-topic distribution in Table 3.1 has higher probabilities for the terms *dog* and *canine* than LDA. These increased probabilities for those terms create the explicit semantic relationship, LDA alone is unable to define.

As discussed in Section 2.5.1 each set of synsets, has a unique index organized into hierarchies. Each hierarchy level expands further to reveal a myriad of synsets, expanding the synsets excessively may cause term frequencies to be incremented unrealistically. Therefore, LDAWN only expands the first level synsets of the hierarchy

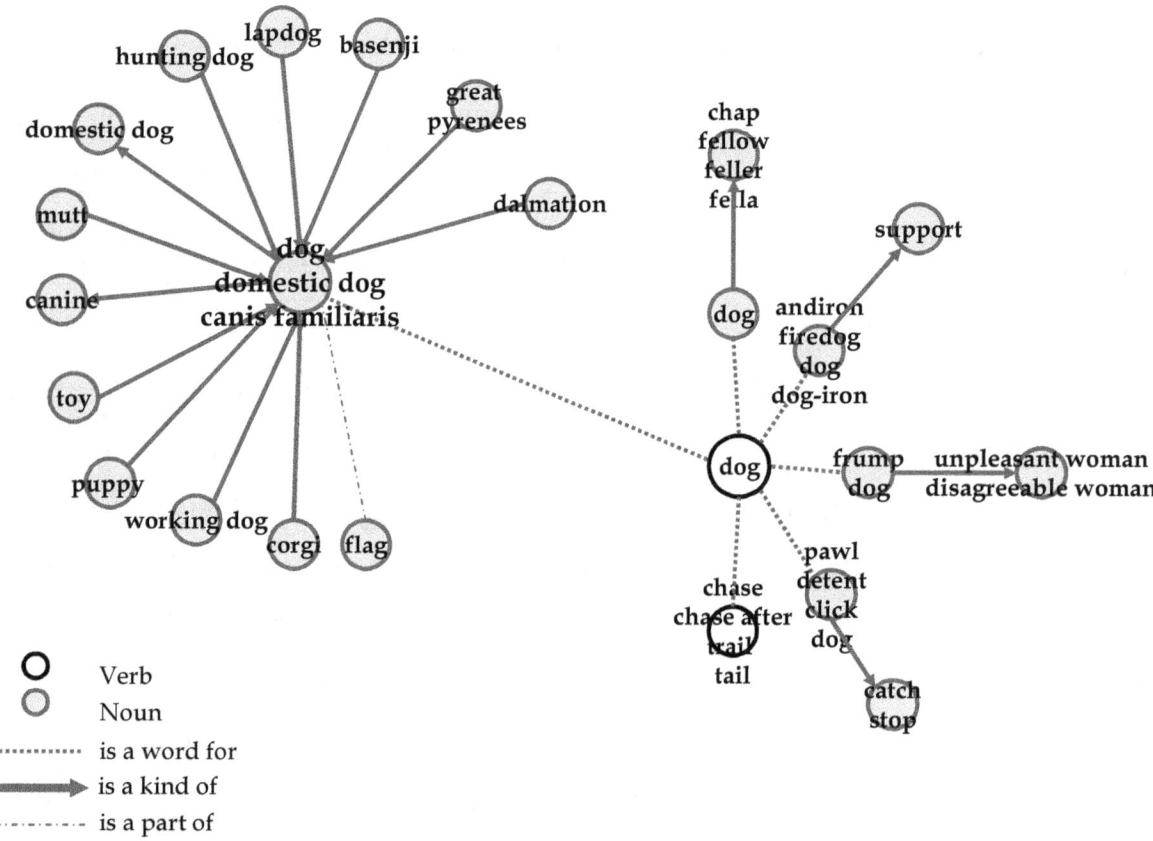

Figure 3.2:   Graphical depiction of WordNet Synsets for *dog* and *canine*.

which avoids traversing too far into the ontology network causing unnecessary computation and unmanageable relationship tracking. This level captures the most prevalent word semantic relationships. However, restricting the number of hierarchy levels could cause LDAWN to overlook important semantic relationships. Further study can investigate if additional levels yields better results. Figure 3.2 is a graphical depiction of the first level synsets for the terms *dog* and *canine*.

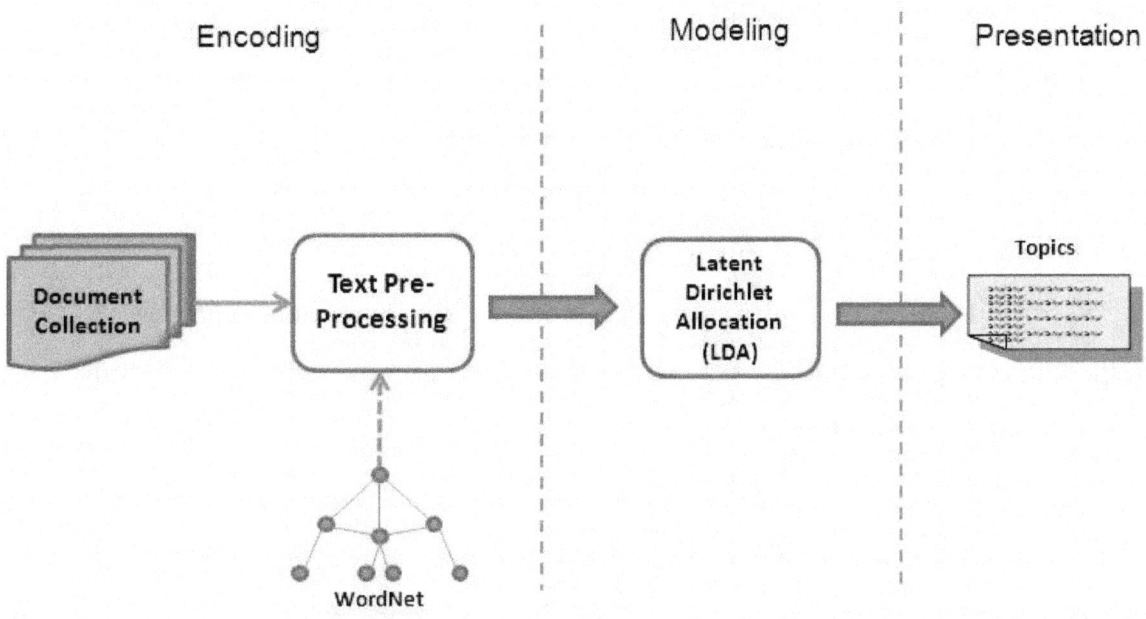

Figure 3.3:   The LDAWN Process.

The LDAWN process is shown in Figure 3.3 and described in the following steps:

1. Parse and remove stopwords (stemming not used).

2. Store documents and terms in the collection.

3. Parse and build vocabulary.

4. Pre-process and encode data as term frequencies.

5. Find semantically related terms in the vocabulary and weight them using Word-Net.

6. Use repeated Gibbs sampling and LDA algorithm for 200 iterations or until convergence.

7. Output the per-document topic distribution θ and per-topic word distribution ϕ.

3.2.3 Experimental Design.   LDA and LDAWN are trained on four text corpus to compare the generalization performance of these models. The first collection is 20 Newsgroups which is a collection of pre-categorized newsgroups into 20 topic areas [29]. OHSUMED is a set of references from MEDLINE, classified into topics where documents can belong to multiple topics. MEDLINE is an on-line medical information database, consisting of titles and abstracts from 270 medical journals over a 5-year period from 1987-1991 [19]. Also, the NIPS collection are the abstracts from the "Neural Information Processing Systems" conference containing the abstracts of the submitted papers over a 5-year period from 2000-2005 and are unlabeled [14]. Finally, experiments are conducted using a collection of unlabeled classified improvised explosive device (IED) reports.

Using the four corpora listed above, 90% of the documents are trained for each data set on LDA and LDAWN models. LDA is allowed to run up to 200 iterations or until convergence, as discussed in Blei et al. [7]. The model parameters taken from the output during training are fixed and used on the test set. These inputs include the topic-distribution θ, α, and β. According to Steyvers and Griffiths, good values for α, and β are $\alpha=50/T$ and $\beta=0.01$ based on the number of topics T and the vocabulary size [41]. Keeping θ learned from the training set, the remaining documents are used as a test set to calculate the perplexity for specified number of topics. The models are evaluated for 10 topic values from 10 to 100 in increments of 10. Figure 3.4 depicts the experiment design with LDA at the top of the figure and LDAWN at the bottom.

Further experiments are conducted on the four collections using independent and pseudorandom training and testing sets for each experiment. There are a total of five experiments per collection to include previous experiments. As before 90% of the collection used in training and 10%held out for testing. This is to ensure collection biasing is avoided. To find the best model parameters additional experiments are conducted where the values of α and β and varying topic numbers are explored to see how the LDA and LDAWN models react and if document modeling can be further improved. Experiments on the 20 Newsgroups and IED collections are conducted with $\alpha=50/T$ at topics numbers 50, 100 and 200. The β parameter is varied from 0.01, 0.02 and 0.05.

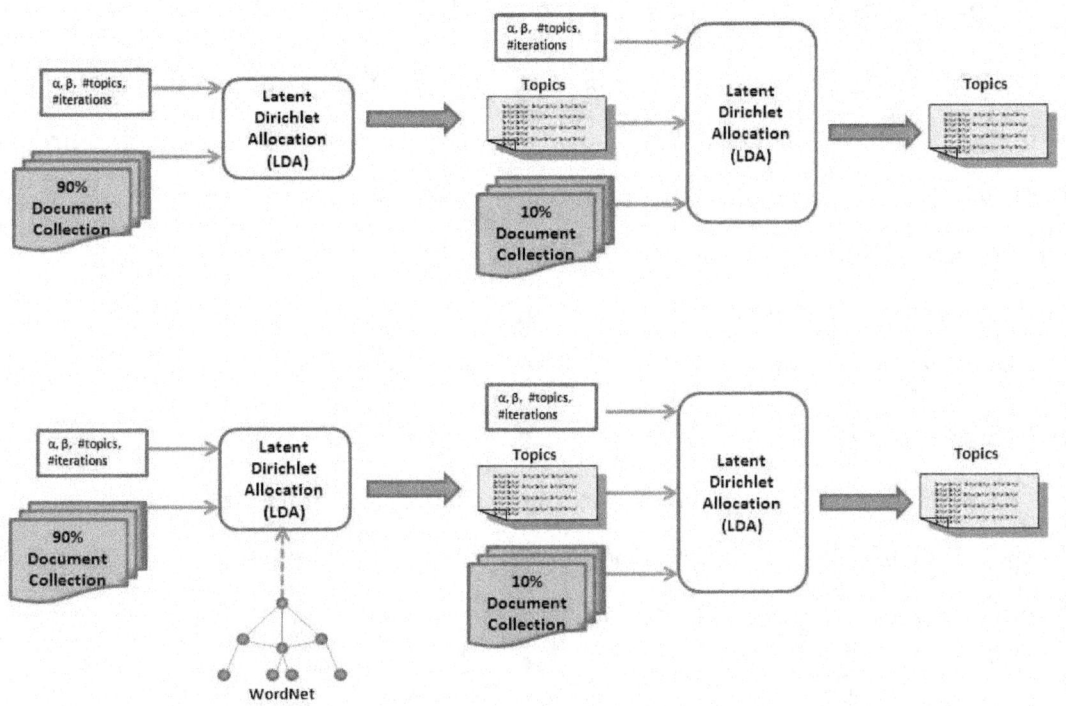

Figure 3.4:   Experimental Design for LDA (Top) and LDAWN (Bottom).

The evaluation metric perplexity, see Equation 2.37 in Chapter 2, a common metric for evaluating natural language processing models [25], are used to evaluate the models. The perplexity value computed on the held-out test data indicates how well the model is able to generalize unseen data. The lower the perplexity the better the model is able to generalize. Four experiments were conducted to thoroughly compare the performance of LDAWN against that of LDA. Experiment one consisted of training and testing over all four copora. Experiment two consisted of obscuring two of the collections from their pre-categorized state, OHSUMED and 20 Newsgroups collections. Experiment three is the mean perplexity values over all experiments and include standard deviation error bars. Experiment four explored the results of adjusting the prior parameters α and β for the per-document topic distribution θ and the per-topic word distribution ϕ.

3.3 Rank Threshold Detection with LDAWN

LDA-SOM clusters documents based on the self-organizing map after document modeling with the Latent Dirichlet Allocation algorithm and topic selection are complete. LDA-SOM leverages the word-topic distribution output of LDA to produce a visualization of those document clusters. The LDA-SOM process is similar to the LDAWN process in that initial preprocessing of documents are parsed and stopwords are removed, with no stemming. The vocabulary for the collection is created and data is encoded as term frequencies in a term-document matrix.

The documents are ranked according to their relevance to a given query by combining the Dirichlet smoothing document model $P(w|D)$ with the LDA model [48]. This combination is given by the following equation:

$$P(w|D) = \lambda \left(\frac{N_d}{N_d + \mu} P_{ML}(w|D) + (1 - \frac{N_d}{N_d + \mu}) P_{ML}(w|coll) \right) + (1 - \lambda) P_{lda} w|D$$

(3.1)

where P_{ML} is the probability from original document model and P_{lda} is the probability from the LDA model. The parameters λ and μ are set at μ=1000 and λ=0.7, which achieve the best results according to Wei et al. [48]. The hybrid probabilistic query model differs in one area, $P_{ML}(w|coll)$ is changed to $P_{ML}(w|C)$, where C is the cluster containing the document [36].

$$P(w|D) = \lambda \left(\frac{N_d}{N_d + \mu} P_{ML}(w|D) + (1 - \frac{N_d}{N_d + \mu}) P_{ML}(w|C) \right) + (1 - \lambda) P_{lda} w|D \quad (3.2)$$

This change gives the retrieval process a distinct advantage of assigning probability to document that are relevant to the query in which the query terms to not explicitly appear in the documents [36]. The following steps and Figure 3.5 depict the LDA-SOM process:

1. Use LDA to classify the words and documents into topics.

2. Look at the topics that emerge and decide which of the topics are relevant to the the user.

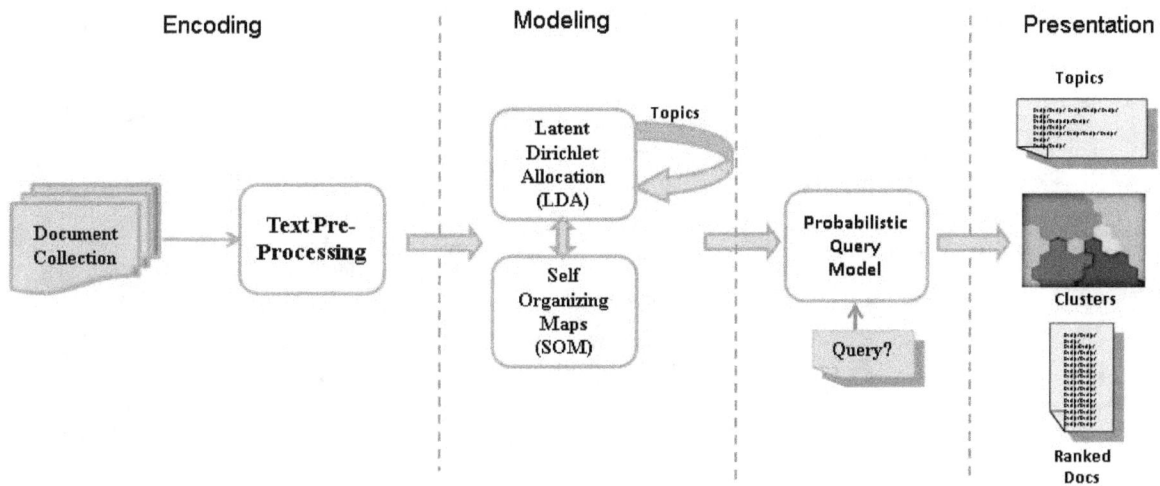

Figure 3.5:   LDA-SOM IR Process

3. Take the documents and put them in a SOM using the probabilities for the relevant topics as the dimensions of the data.

4. Cluster the SOM and determine the largest cluster. This is the one that has low probability for all dimensions (i.e., topics).

5. Discard the documents associated with the largest cluster.

6. Take the remaining documents and run LDA on them to generate new topics.

7. Run SOM using the remaining documents and the new topics.

8. Cluster the SOM.

9. Use the clusters and the LDA topics to rank documents to user defined queries using the hybrid probabilistic query model Equation 3.2.

3.3.1 Threshold Problem Definition.   Previous work by Millar [36], LDA-SOM, rank document relevance to a query independent of whether or not the query terms appeared in the document. More importantly, his implementation overlooks query terms that do not explicitly co-occur and is discounted by LDA. In LDAWN, WordNet is used to supplement the rank documents based on the query terms including any of its synsets to leverage terms that co-occur. In addition, since all remaining

47

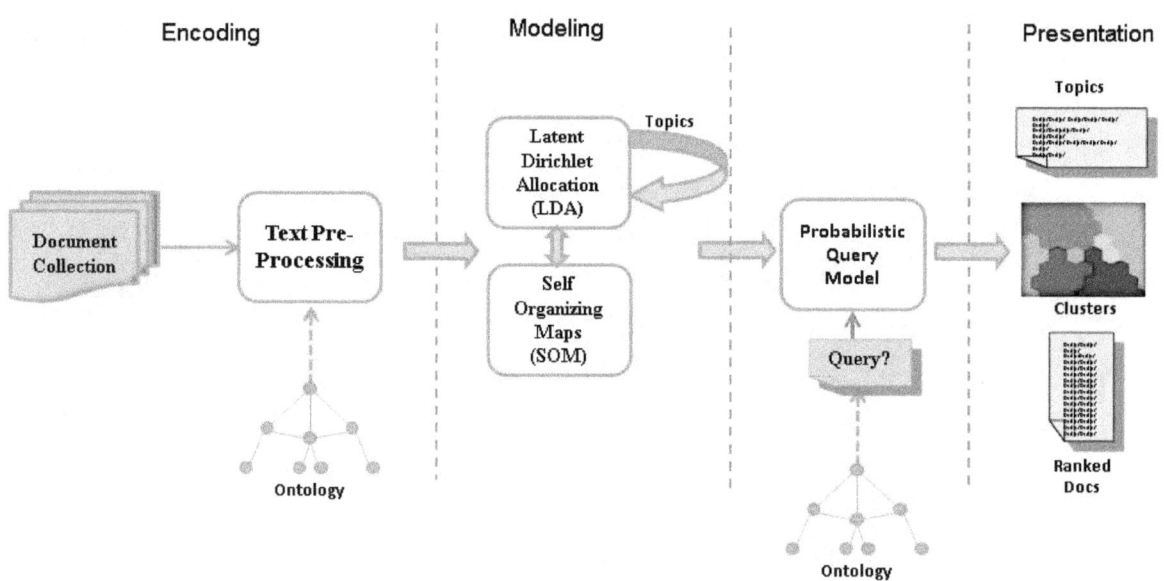

Figure 3.6:   LDAWN IR Process

documents are ranked and returned to the user a rank threshold of relevancy should be automatically defined.

3.3.2 Threshold Goals and Hypothesis.   Equation 3.2 bases the relevancy of documents to a query independently of whether the query terms appear in a document or not, therefore WordNet can be leveraged by finding those documents that the query terms appear and any of its WordNet synsets. Also, a rank threshold can be automatically detected by determining the point at which the query terms or their synsets no longer appear in the documents. This provides the user a point at which documents are no longer relevant without searching the entire ranked documents list. The LDAWN IR process is depicted in Figure 3.6 the only difference from the LDA-SOM IR process is the incorporation of ontologies.

Experiments were conducted on the 20 Newsgroups collection using Millar's LDA-SOM algorithm and LDAWN algorithm with equation 3.2 and the following parameters, $\alpha=50/T$, $\beta=0.01$, $\mu=1000$ and $\lambda=1000$. Both algorithm's are allowed to run 200 iterations and the ranked document list contains the automatic threshold, indicated by a T in the document list. The comparison metric is precision and recall, where the 20 Newsgroups are in categories and the query can be assigned to a specific category, therefore can be treated as a semi-labeled collection.

IV. Results and Analysis of LDAWN and Rank Threshold Detection

4.1 LDAWN Proof of Concept

Test results for the first experiment compared the four collection perplexity values, see Figures 4.1 through 4.4, show that LDAWN garnered less (better) perplexity values in a great majority of topic values. Figure 4.1 is the 20 Newsgroups collection, which is pre-categorized into 20 topic areas. Pre-categorization can be inferred since the LDA and LDAWN models are consistently similar in their perplexity values at each topic increment. In Figure 4.2 the perplexity values for the LDAWN model are lower at each topic increment than the LDA model, which means the LDAWN model is able to generalize unseen data better than the LDA model alone. Figure 4.3 are the OHSUMED collection, here the documents have been obscured from their labeled topics. The perplexity values on the training set and again the LDAWN model has a lower perplexity at each topic increment. In Figure 4.4, the IED reports collection prove LDAWN is able to generalize unseen data better than LDA in the majority of test cases. However, in the IED collection at topic numbers 30 and below LDAWN's perplexity increases dramatically which could be due to the held-out set composition. If the test set contained a higher number of words that did not appear in the training set and WordNet did not find their synsets, their probabilities would be lower thereby increasing the perplexity.

The overall improvement, reduction in perplexity, at 100 topics are 9.8% for 20 Newsgroups, 19% for NIPS, 15% for OHSUMED, and 28% for IED. The results for the NIPS, OHSUMED, and IED collections show that the LDAWN model when faced with a previously unseen document which may contain words that did not appear in the training documents are able to generalize those words better than the LDA model. These words most likely have smaller probabilities which make the perplexity of the unseen documents increase in the LDA model. Since the LDAWN model is able to find semantically related words in these documents, those word probabilities increase which decrease the perplexity for those unseen documents. However, as seen in the

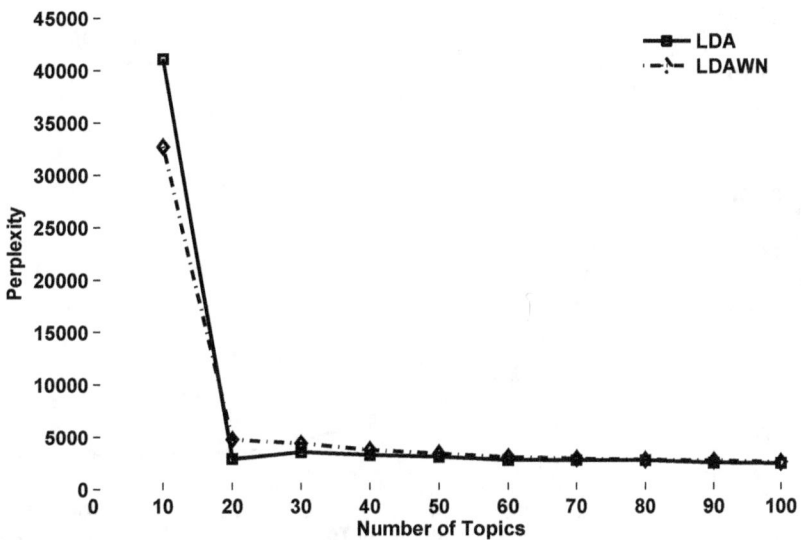

Figure 4.1:   Perplexity results on the 20 Newsgroups collection.

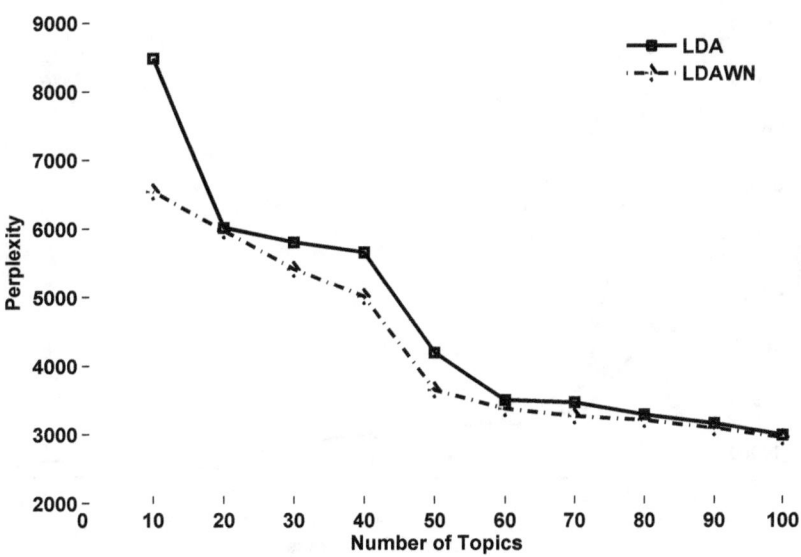

Figure 4.2:   Perplexity results on the NIPS collection.

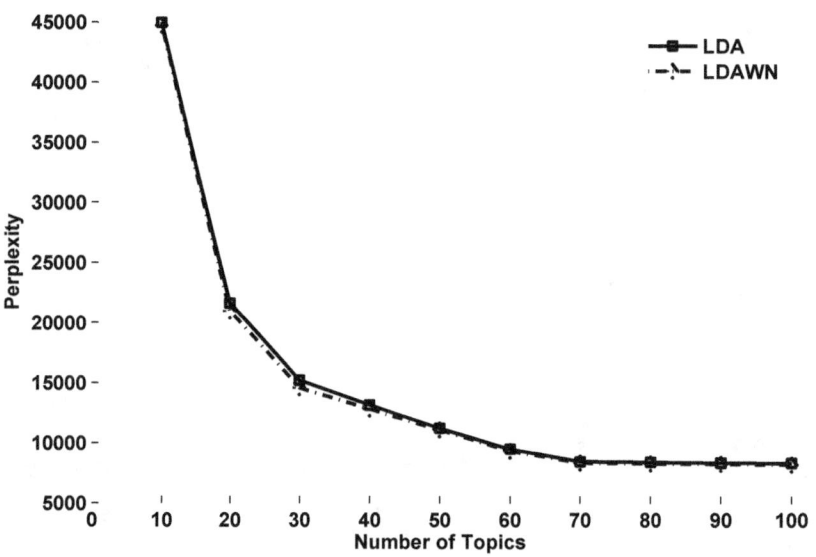

Figure 4.3: Perplexity results on the OHSUMED collection.

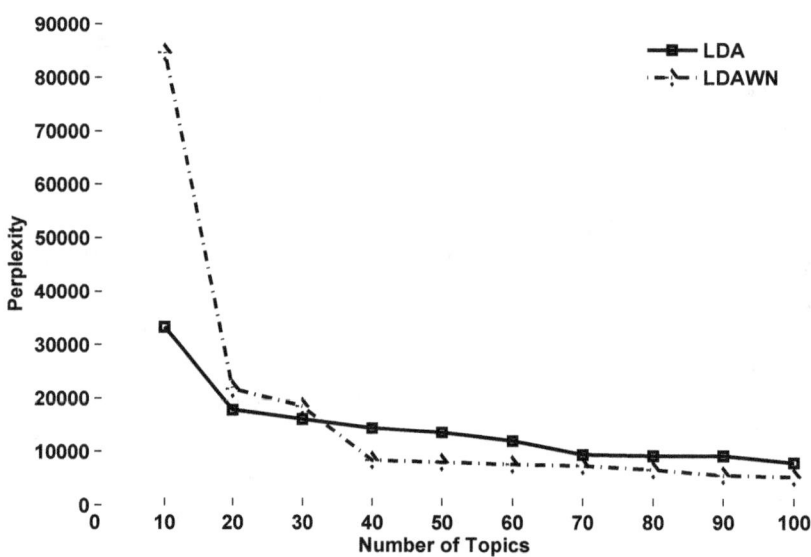

Figure 4.4: Perplexity results on the IED collection.

Table 4.1:   LDA and LDAWN Models Training Times.

	20 Newsgroups (33 MB)	OHSUMED (60.1 MB)	NIPS (35.8 MB)	IED (12.2 MB)
LDA	37 min	195 min	43 min	64 min
LDAWN	110 min	250 min	94 min	137 min

IED collection at topic numbers 30 and below, LDAWN's perplexity indicates that there are a number of terms that do not have synsets and do not occur frequently and therefore have low probabilities. This could also be a collection anomaly requiring further testing.

A drawback of using the LDAWN model for document modeling is the increased runtime in searching through the synsets and incorporating the additional words. Our experiments show that it takes approximately twice as long to run the LDAWN model. Table 1 shows the associated runtime for each collection with the corresponding model. These runtimes are also associated with the size of the collection. The OHSUMED collection is the largest of the three and therefore takes the longest to run for both models. System memory for these tests is 3 GB of RAM and a 2.7 GHZ processor, increasing system memory and processor speed may reduce the training runtime.

4.1.1   Analysis.   These experiments show significant improvements over previous work using LDA to model documents by incorporating the WordNet ontology to help uncover hidden semantic relationships. For any given document, we incremented term frequencies for all terms in the document matching terms in the synsets of WordNet. Then, we incorporated this enhanced term-document matrix into the LDA model to compute the topic distribution. LDA estimated the per-document topic distribution and per-topic word distribution and output the probabilities for each topic distribution.

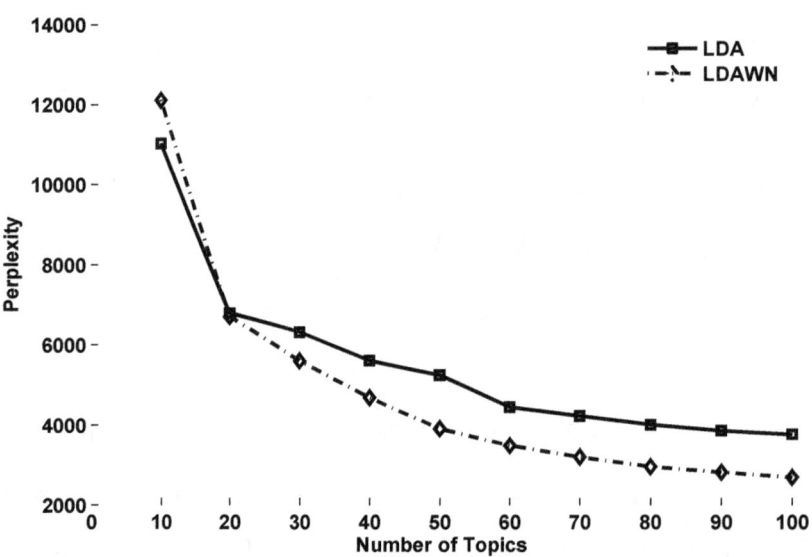

Figure 4.5: Perplexity results on the 20 Newsgroups obscured collection.

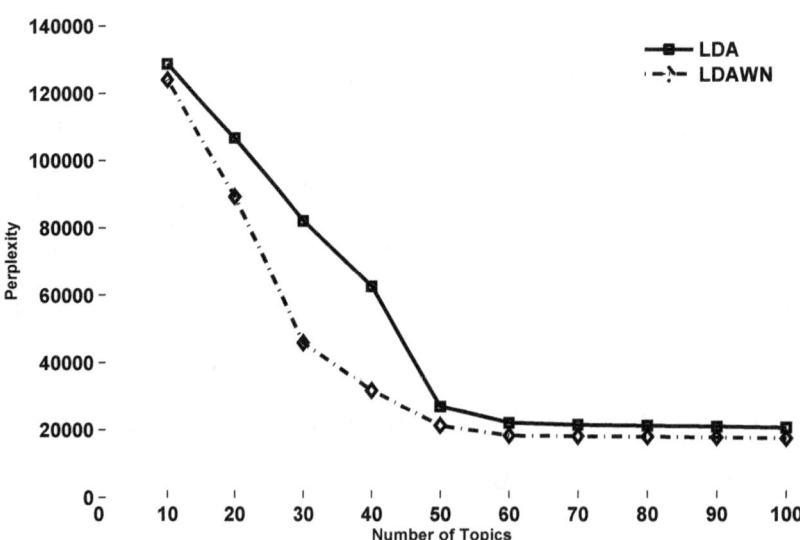

Figure 4.6: Perplexity results on the OHSUMED obscured collection.

4.2 *LDAWN Randomized Collections*

A majority of each collection are trained using the LDAWN model to classify the documents into topics. The 10% held-out set is used to test and measure the perplexity, using perplexity Equation 2.37 as discussed in Chapter 2, of each collection for several numbers of topics. Results show that augmentation of LDAWN, fared better than LDA alone, i.e., LDAWN achieves a better generalization of documents in each collection.

Test results for experiment two are the obscured (mixed-up) 20 Newsgroups and OHSUMED collections and test if the collection categorization affect perplexity values, see Figures 4.5 and 4.6. Figure 4.5 is the 20 Newsgroups collection, where the documents were obscured from their pre-categorized topics. LDAWN outperforms LDA at all topic values in the obscured collection where the perplexity's were similar along most topic values in experiment one. Figure 4.6 is the obscured OHSUMED collection where the documents are not in their predefined state as classified topics. Again, LDAWN outperforms LDA at all topic values in the obscured collection where the perplexity's were similar along most topic values in experiment one.

4.2.1 Analysis. During this experiment the results for the OHSUMED and 20 Newsgroups collections are different from the topic categorized findings in experiment one. These results are due to obscuring the OHSUMED and 20 Newsgroups collection from their original topic category's. This experiment shows that collections that are labeled or pre-categorized pose similar perplexity's at all topic values for LDAWN and LDA although LDAWN's values are slightly lower. This fact tells us that the LDAWN model it most helpful when collections are not categorized or labeled, which is the case when document and topic modeling are the most useful.

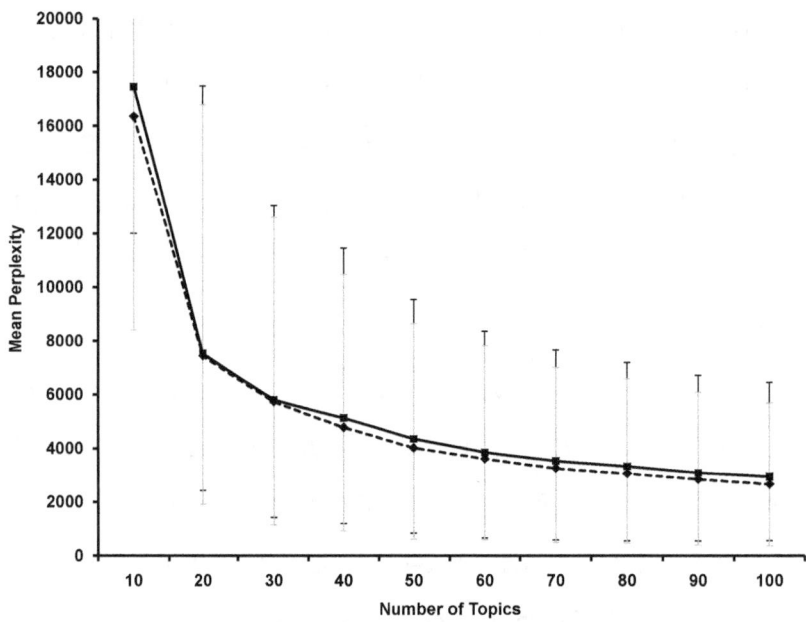

Figure 4.7:   Mean Perplexity results for the 20 Newsgroups collection.

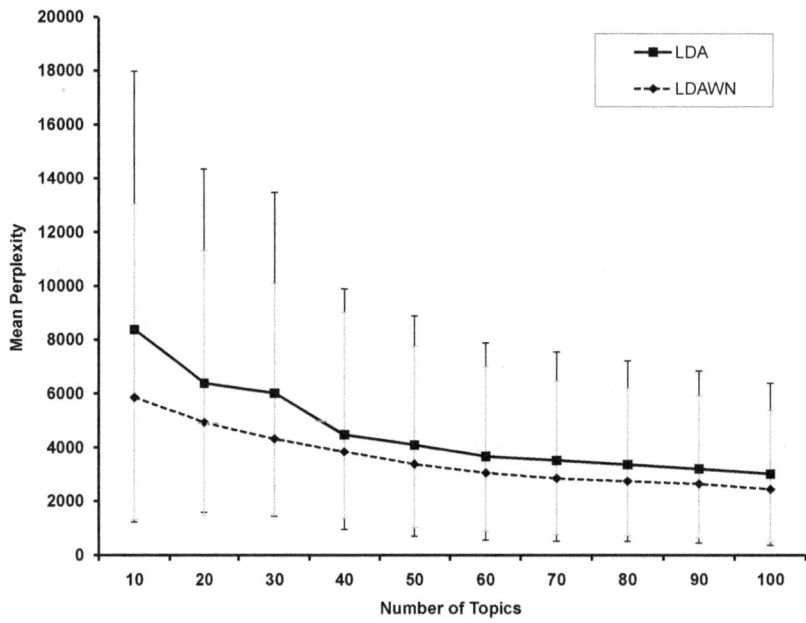

Figure 4.8:   Mean Perplexity results for the NIPS collection.

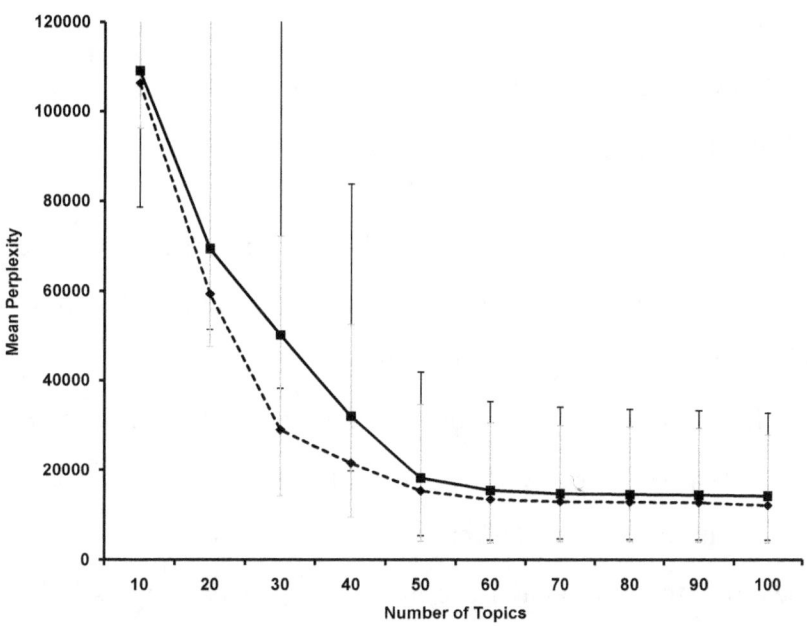

Figure 4.9:   Mean Perplexity results for the OHSUMED collection.

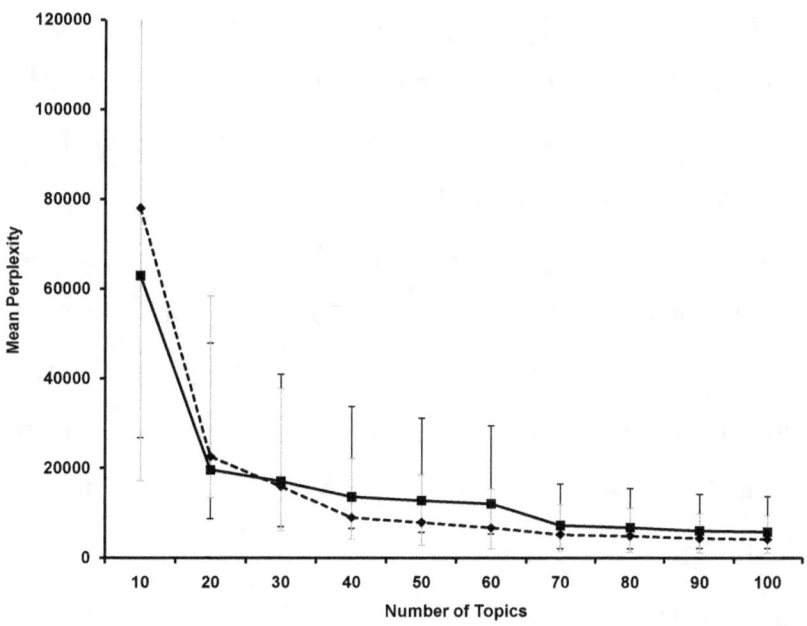

Figure 4.10:   Mean Perplexity results for the IED collection.

4.3 LDAWN Confidence Testing

Experiment three mean perplexity values over five experiments on all four collections, see Figures 4.7 through 4.10, shows the mean perplexity values for the 20 Newsgroups, NIPS, OHSUMED and collections.

4.3.1 Analysis.   Test results over the five experiments per collection with the perplexity mean and standard deviation, shown in Figures 4.7 through 4.10, show that LDAWN garnered less (better) perplexity values in a great majority of topic values. Figure 4.7, is the 20 Newsgroups collection, which is pre-categorized into 20 topic areas. This pre-categorization can be inferred since the LDA and LDAWN models are consistently similar in their perplexity values at each topic increment, meaning that the documents in each pre-labeled topics categories belong in similarly inferred topics. Overall, the perplexity values for the LDAWN model are lower at each topic increment than the LDA model, which means the LDAWN model is able to categorize the new documents better than the LDA model alone. Figure 4.8, is the NIPS collection, LDAWN consistently outperforms LDA at every topic number. Figure 4.9 bottom left, is the OHSUMED collection perplexity values on the training set and again the LDAWN model has a lower perplexity at each topic increment. In Figure 4.10, the IED reports collection prove, once again, that LDAWN is able to generalize unseen data better than LDA in the majority of test cases. However, in the IED collection at topic numbers 30 and below, LDAWN's perplexity increases dramatically which could be due to the held-out set composition. If the test set contained a higher number of words that did not appear in the training set and WordNet did not find their corresponding synsets, their probabilities would be lower thereby increasing the perplexity. This can occur at a lower number of topics when the collection is diverse.

The results for the NIPS, OHSUMED, and IED collections show that the LDAWN model when faced with a previously unseen document, which may contain words that did not appear in the training documents, are able to generalize those words better than the LDA model. These words most likely have smaller probabilities which result in the perplexity of the unseen documents to increase in the LDA model. Since the LDAWN model is able to find semantically related words in these documents, those word probabilities increase which decrease the perplexity for those unseen documents. However, as seen in the IED collection at topic numbers 30 and below, LDAWN's perplexity indicates that there are a number of terms that do not have synsets and do not occur frequently and therefore have low probabilities. This may also be a collection anomaly requiring further testing.

Figures 4.7 through 4.10 also shows the standard deviation depicted by the error bars, black for plain LDA and gray for LDAWN, among the four collections over five experiments. The variances at 10 topics consistently have a large spread, on the OHSUMED, 20 Newsgroups and IED collections, due to more outliers or increased perplexities at low topic numbers. This is reasonable because it is more difficult to construct topic models at low topic numbers because the document model is generated by first picking word distributions over the topics. Therefore, if the number of topics is low then the word distribution is severely restricted causing a high variance. Alternately, LDAWN has a smaller variance than LDA in all the collections.

4.4 Determining Best α and β Parameters

To empirically determine the best α and β to use, experiments on the 20 Newsgroups and IED collections are conducted with $\alpha=50/T$ at topics numbers 50, 100 and 200. The β parameter is varied from 0.01, 0.02 and 0.05 depicted in Figures 4.11 through 4.13. Figure 4.11 is the 20 Newsgroups collection, at 50 topics, top left, with $\beta=0.01$ the LDAWN perplexity decreases drastically as it approaches 50 topics as well as LDAWN outperforming LDA at most topic numbers. This is also true with LDAWN and 100 topics Figure 4.12, with parameter $\beta=0.01$ perplexity steadily decreases and is lower than the perplexity at $\beta=0.02$ and 0.05. LDAWN still outperforms LDA at each topic number. Figure 4.13 at 200 topics and $\beta=0.02$, there is a

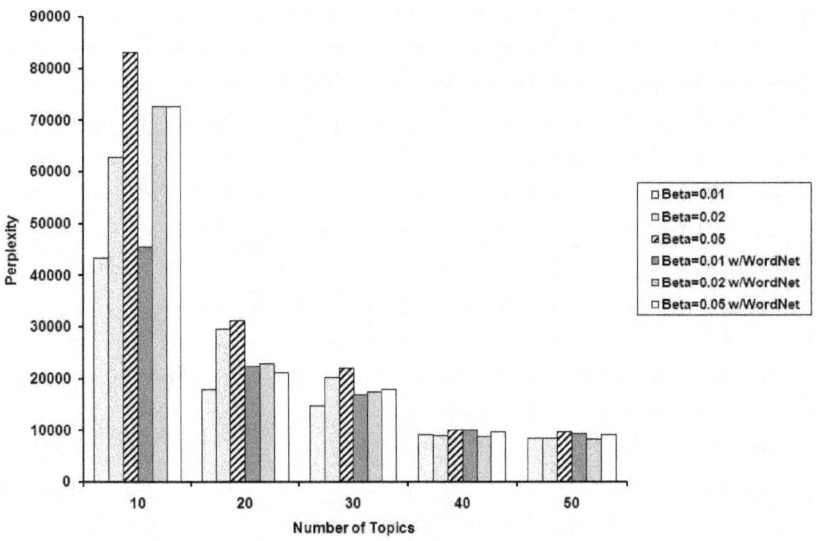

Figure 4.11:   20 Newsgroups 50 Topics at $\alpha = 50/T$ and $\beta = 0.01$, 0.02, 0.05.

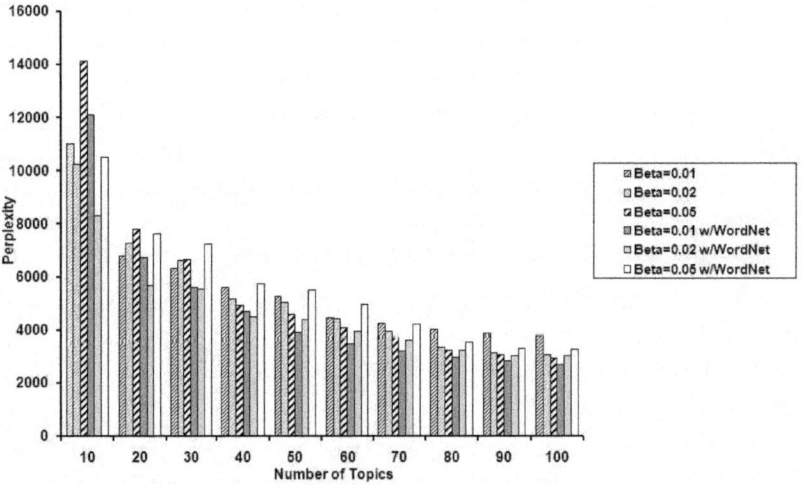

Figure 4.12:   20 Newsgroups 100 Topics at $\alpha = 50/T$ and $\beta = 0.01$, 0.02, 0.05.

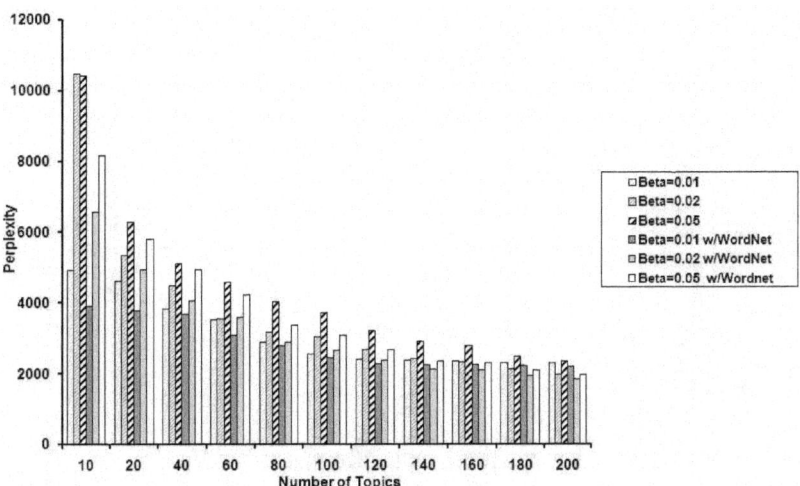

Figure 4.13:   20 Newsgroups 200 Topics at α=50/T and β=0.01, 0.02, 0.05.

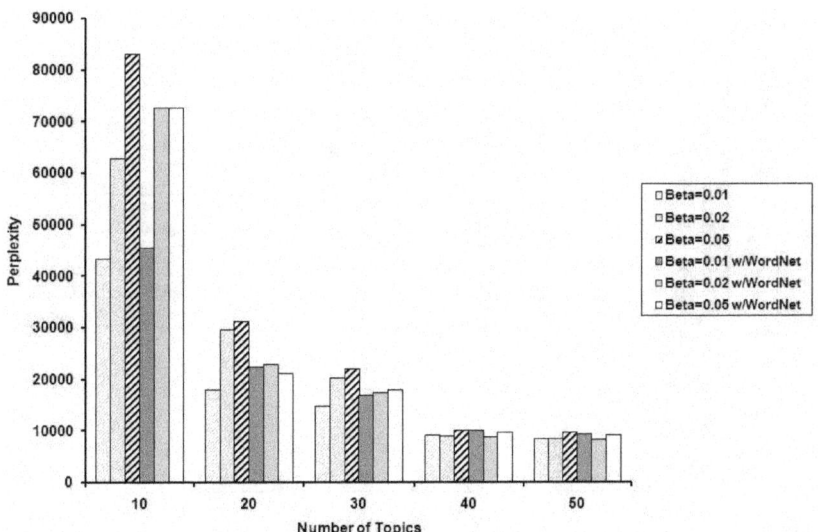

Figure 4.14:   IED 50 Topics at α=50/T and β=0.01, 0.02, 0.05.

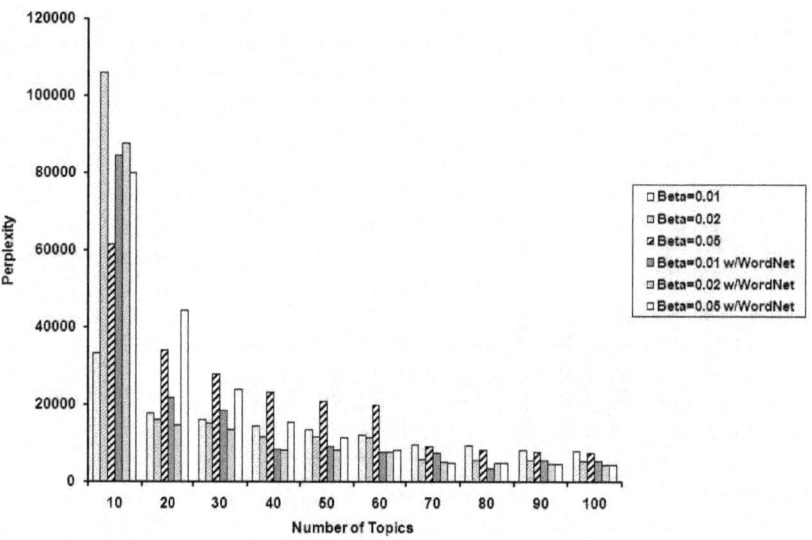

Figure 4.15:   IED 100 Topics at $\alpha=50/T$ and $\beta=0.01$, 0.02, 0.05.

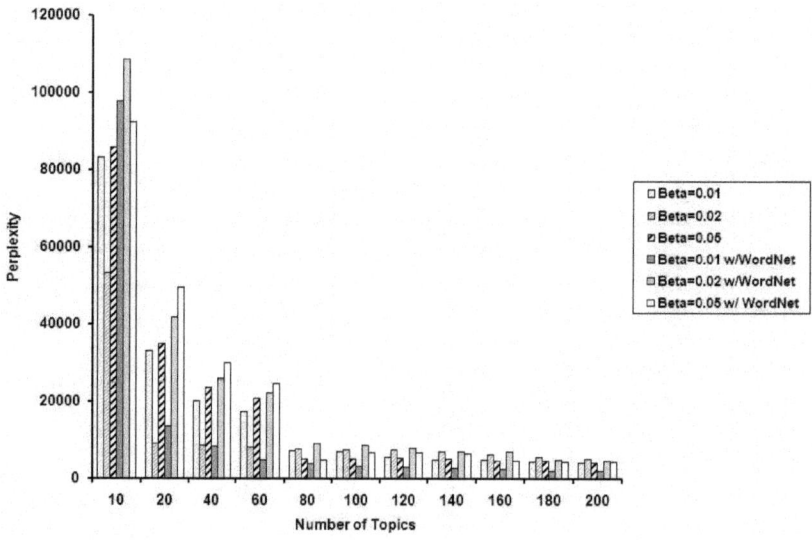

Figure 4.16:   IED 200 Topics at $\alpha=50/T$ and $\beta=0.01$, 0.02, 0.05.

noticeable decline in the perplexity. LDAWN consistently has a lower perplexity than LDA with β=0.01 and α=0.25 but at larger topic numbers such as 200, better results are achieved with β=0.02.

As with the 20 Newsgroups, the IED collection showed similar results with the varying values for α and β. Figure 4.14 through 4.16 is the IED collection, at 50 topics, Figure 4.14, with β=0.02 the LDAWN perplexity decreases as it approaches 50 topics as well as LDAWN outperforming LDA at most topic numbers. This is also true with LDAWN at 100 topics Figure 4.15, with parameter β=0.02 perplexity steadily decreases and is lower than the perplexity with β=0.01 and 0.05. LDAWN still outperforms LDA at each topic number. Figure 4.16 with 200 topics and β=0.01, there is a noticeable decline in the perplexity. LDAWN consistently has a lower perplexity than LDA with β=0.02 and α=0.25 but at larger topic numbers such as 200, better results are achieved with β=0.01.

4.4.1 Analysis.   These experiments were designed to test the best parameters for the document models. As proposed by Steyvers and Griffiths [41], we also found the values of α=50/T and β=0.01 produced the best overall results. However, these values are best fit for an unlabeled collection like the IED collection. When the model is faced with a categorized/semi-labeled collection such as 20 Newsgroups β=0.02 fared better with large topic numbers. So when determining the best parameters, the collection and desired number of topics should be considered when choosing α and β.

4.5   *Rank Threshold Detection Using WordNet*

LDAWN also incorporates a query model for information retrieval purposes. The documents are ranked according to their relevance to a given query by combining the Dirichlet smoothing document model with the LDA model as proposed by Wei and Croft [48].

4.5.1 LDAWN Threshold Detection Results. Table 4.2 are the LDAWN top 50 results on the query term *bike*. The value of *true* is given to a document that contains the query term or its WordNet sysnet. Notice that a value of *true* does not appear until three quarter of the way down the list. This is unusual since most IR system results would rank the documents with the query term much higher. In order to increase the score of the documents that contain the query term a multiplier must be used. Therefore, a multiplier is applied to the rank score of each document in the ranked list that contains the query term or its WordNet synset. The multiplier moves the documents that contain the query term or its WordNet synset closer to the top, these results are shown in Table 4.3. After the multiplier is applied the last *true* value is found and the threshold is drawn as seen in Table 4.4, these are ranked 650-700. The threshold was found at document 1441 indicated by the *T* for the suggested threshold, at rank 653, which is about the halfway point of the total 1497 documents.

To further evaluate the performance of LDAWN, precision and recall metrics are calculated for both LDA and LDAWN. The OHSUMED collection was used since it includes labeled queries, i.e., subject matter experts determined which documents in the collection are relevant to the queries. Table 4.5 and 4.6 are the precision and recall results from the OHSUMED collection for labeled query's Q1 through Q15. Note: Q8 and Q14 have no relevant documents.

Figure 4.17 depicts the average recall versus precision at 11 levels of recall for the two algorithms averaged across the 13 queries. Like Figure 4.17, typically these graphs slope downward from left to right, enforcing the notion that as more relevant documents are retrieved (recall increases), the more non-relevant documents are retrieved (precision decreases). Therefore curves closer to the the upper right corner of the graph, e.g., closest to 100% precision and recall, perform better. Since LDAWN garnered higher precision and recall for every query over LDA and is closer to the upper right corner, LDAWN's performance is superior to LDA. Therefore, we conclude that incorporating WordNet into the query process is also beneficial for information retrieval performance.

64

Table 4.2:   Top 50 Results for *bike* LDAWN without multiplier.

Rank ID	Doc ID	Score	Contains Query Term
1	871	0.54251	FALSE
2	1167	0.09532	FALSE
3	920	0.09036	FALSE
4	1154	0.07438	FALSE
5	697	0.06602	FALSE
6	867	0.06386	FALSE
7	1270	0.06045	FALSE
8	997	0.05942	FALSE
9	1309	0.05517	FALSE
10	1022	0.05467	FALSE
11	1139	0.05178	FALSE
12	771	0.04916	FALSE
13	1122	0.04853	FALSE
14	776	0.04793	FALSE
15	1318	0.04666	FALSE
16	842	0.0449	FALSE
17	775	0.04424	FALSE
18	738	0.04328	FALSE
19	1308	0.04245	FALSE
20	687	0.04235	FALSE
21	820	0.04221	FALSE
22	747	0.04191	FALSE
23	992	0.04188	FALSE
24	843	0.04149	FALSE
25	1267	0.04144	FALSE
26	792	0.04059	FALSE
27	1096	0.04058	FALSE
28	1260	0.0404	FALSE
29	817	0.03958	FALSE
30	1104	0.03955	FALSE
31	824	0.03878	FALSE
32	823	0.03875	FALSE
33	1358	0.03836	FALSE
34	1231	0.38316	TRUE
35	934	0.03808	FALSE
36	1381	0.03806	FALSE
37	1249	0.03765	FALSE
38	1243	0.03712	FALSE
39	804	0.03707	FALSE
40	1197	0.03674	FALSE
41	922	0.03673	FALSE
42	835	0.03579	FALSE
43	650	0.03506	FALSE
44	998	0.035	FALSE
45	1087	0.03484	FALSE
46	1110	0.03474	FALSE
47	1210	0.03472	FALSE
48	1129	0.03449	FALSE
49	972	0.03448	FALSE
50	1153	0.03447	FALSE

Table 4.3: Top 50 Results for *bike* LDAWN with multiplier.

Rank ID	Doc ID	Score	Contains Query Term
1	871	0.5425	FALSE
2	1231	0.3832	TRUE
3	1268	0.3176	TRUE
4	813	0.3052	TRUE
5	1072	0.2189	TRUE
6	1380	0.2172	TRUE
7	1123	0.1942	TRUE
8	1186	0.1879	TRUE
9	700	0.174	TRUE
10	1182	0.155	TRUE
11	760	0.1452	TRUE
12	644	0.1432	TRUE
13	1325	0.1396	TRUE
14	1097	0.1378	TRUE
15	1208	0.1354	TRUE
16	77	0.1217	TRUE
17	1167	0.0953	FALSE
18	920	0.0904	FALSE
19	965	0.0833	TRUE
20	942	0.0769	TRUE
21	757	0.0748	TRUE
22	1154	0.0744	FALSE
23	1200	0.0729	TRUE
24	913	0.0724	TRUE
25	1328	0.0723	TRUE
26	41	0.0719	TRUE
27	725	0.0704	TRUE
28	1058	0.07	TRUE
29	641	0.0697	TRUE
30	692	0.0695	TRUE
31	380	0.068	TRUE
32	89	0.0679	TRUE
33	1173	0.067	TRUE
34	978	0.0663	TRUE
35	697	0.066	FALSE
36	359	0.0657	TRUE
37	1491	0.0656	TRUE
38	417	0.0655	TRUE
39	1434	0.0655	TRUE
40	120	0.0653	TRUE
41	1404	0.0651	TRUE
42	987	0.0643	TRUE
43	936	0.0642	TRUE
44	1245	0.064	TRUE
45	867	0.0639	FALSE
46	308	0.0638	TRUE
47	1001	0.0633	TRUE
48	698	0.0631	TRUE
49	52	0.0624	TRUE
50	424	0.0623	TRUE

Table 4.4:   50 Results for *bike* LDAWN that include threshold.

Rank ID	Doc ID	Score	Contains Query Term
632	238	0.01939	TRUE
633	1161	0.01939	FALSE
634	140	0.01939	TRUE
635	44	0.01938	TRUE
636	248	0.01937	TRUE
637	741	0.01936	FALSE
638	955	0.01936	FALSE
639	251	0.01935	TRUE
640	422	0.01935	TRUE
641	97	0.01934	TRUE
642	1367	0.01923	FALSE
643	1130	0.01919	FALSE
644	389	0.01914	TRUE
645	94	0.01911	TRUE
646	1215	0.01908	FALSE
647	961	0.01905	FALSE
648	962	0.01896	FALSE
649	98	0.01894	TRUE
650	854	0.01889	FALSE
651	743	0.01888	FALSE
652	695	0.01884	FALSE
T 653	1441	0.01881	TRUE
654	796	0.0188	FALSE
655	1351	0.01867	FALSE
656	1410	0.01867	FALSE
657	1116	0.0186	FALSE
658	1349	0.01859	FALSE
659	953	0.01856	FALSE
660	1117	0.01852	FALSE
661	1354	0.01851	FALSE
662	1075	0.0185	FALSE
663	1357	0.01837	FALSE

Table 4.5:   LDA-SOM and LDAWN Recall for OHSUMED.

	Q1	Q2	Q3	Q4	Q5	Q6	Q7
LDA-SOM	0.0004	0.0002	0.0015	0.0001	0.00004	0.0006	0.00004
LDAWN	0.0004	0.0002	0.0017	0.0001	0.0002	0.0008	0.00007
	Q9	Q10	Q11	Q12	Q13	Q15	
LDA-SOM	0.0002	0.00	0.00008	0.0001	0.00008	0.0002	
LDAWN	0.0002	0.00004	0.00008	0.0001	0.0001	0.0002	

Table 4.6:   LDA-SOM and LDAWN Precision for OHSUMED queries.

	Q1	Q2	Q3	Q4	Q5	Q6	Q7
LDA-SOM	0.64	0.57	0.62	1.00	0.17	0.68	0.50
LDAWN	0.64	0.71	0.80	1.00	0.83	0.95	1.00
	Q9	Q10	Q11	Q12	Q13	Q15	
LDA-SOM	0.57	0.00	0.50	1.00	0.67	0.80	
LDAWN	0.86	0.50	0.50	1.00	1.00	0.80	

4.5.2 Analysis.   This automatic threshold detection method is designed to give the user an estimate of the point at which the documents are no longer relevant to the query. This method is still in its infancy stage but far outperforms a manual binary search of the physical documents. Therefore, there is plenty of room for further explorations and possible improvements. Additional results can be found in Appendix 2, where the threshold is detected for the query terms *battery, motorcycle, tire* and *ride.*

As indicated in the both Tables 4.5 and 4.6 LDAWN has higher precision and recall results for a majority of the queries which means LDAWN retrieves the most relevant documents with respect to the query. Figure 4.17 also shows LDAWN has a higher precision at all 11 levels of recall which validates that LDAWN retrieves more relevant documents than LDA.

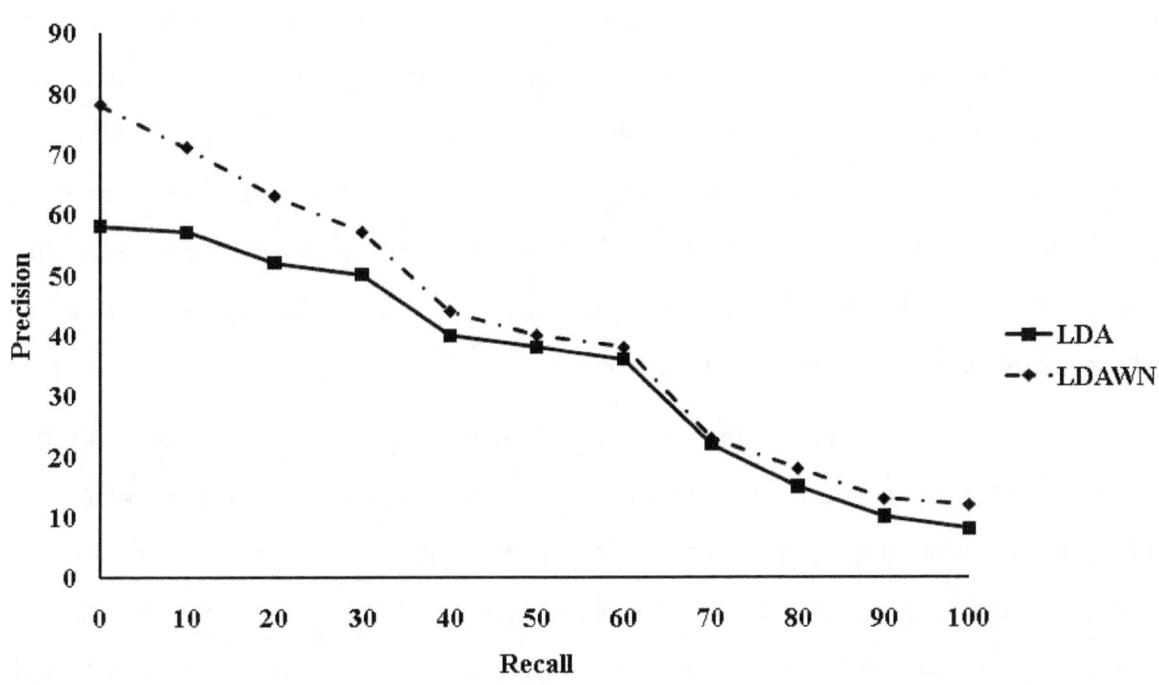

Figure 4.17:   Average precision over 11 levels of recall for OHSUMED queries.

69

V. Conclusions

Latent Dirichlet Allocation (LDA) with WordNet (LDAWN) exposes hidden semantics relationships resulting in improved document modeling, document classification and topic categorization over LDA alone. This technique benefits the e-intelligence/counter-intelligence community by enabling intelligence analysts to quickly extract more relevant information from massive amounts of disparate data, e.g., IED incident reports. For any given document, term frequencies are incremented for all terms in the document with matching terms in WordNet synsets. Then, the resulting term-document matrix is incorporated into the LDA model to compute the topic distribution. LDA estimates the per-document topic distribution and per-topic word distribution and outputs the probabilities for each topic distribution. After unigram classification training over each of the four corpora, a held-out test set is used to measure the perplexity of each collection over several numbers of topics. Our results show that augmentation of LDAWN, fared better than basic LDA, i.e., LDAWN achieves a better generalization of documents in each collection.

The threshold detection method using LDAWN is a way to automatically find a threshold among relevant documents and non-relevant documents in a ranked list. The goal of automation is met which requires no user interaction. In addition, the user has the ability to view documents below the threshold causing no restriction to the user. This method can be used with LDA, LDAWN and other modeling tools that do not have a relevancy threshold detection method.

Results show that augmentation of LDAWN, fared better than basic LDA, i.e., LDAWN achieves a better generalization of documents in the collection by using parameters $\alpha=50/T$ and $\beta=0.01$ as suggested by Steyvers and Griffiths.

5.1 Future Work

Several avenues for future research can further advance this work. Future work includes term reweighting, using domain specific ontologies and further experiments on other labeled collections.

5.1.1 Term reweighting. Term reweighting will modify the term frequencies and therefore modify the probability of the term, increasing the probability of the original term and decreasing the probability of the synset term. This reweighting scheme will be important during the information retrieval process, specifically during document ranking and will give a fair weight to the synset terms. The documents

Currently, LDA-SOM weighs terms equally. To define a distinction between a term and its synset, the synset term and the non-query terms need to be weighted differently. The term weighting process is done during the LDA-SOM term-document matrix generation where the term frequencies are collected. The proposed term weighting scheme would give a term the full weight value of 1.0 if the term is contained in the document. If the term is a synset of the term and both query term and synset terms appears in the document then it would be given a term weight reduction to 0.80. If the term is a synset and appears in the document and the original term does not then the weight would be reduced to 0.75. This term weighting process produces a clear distinction of the term importance, where the original term receives the highest term weight and therefore probability. LDA-SOM experiments for the term weighting process are accomplished with the following parameters $\alpha=50/T$, $\beta=0.01$ and various query's tested on the 20 Newsgroups and IED collections .

5.1.2 Customized Ontologies. WordNet like ontologies can be created and tailored for expected terms in the collection. The tailored ontologies will drastically reduce the ontology size. This reduction in size would improve LDAWN runtime performance and provide domain specific synsets, lowering perplexity. Additionally, OWL (see Section 2.5.2) can be leveraged to generate such customized ontologies.

5.1.3 Evaluate Labeled IED Data. The LDAWN favorable results could be further validated if the IED collection included canned queries and the relevant documents for those queries. Although, very labor intensive process, as the IED collection is a large and dynamic collection, this would provide a validated baseline to ensure LDAWN is the best model to gain insights into this collection. In addition, an

IED subject matter expert should verify the automatic threshold detection process is indeed finding the best possible threshold and if it will be a valuable tool for their analysis.

Appendix A. Stop Word Listing

a	about	after	again	all	almost
also	although	always	among	an	and
another	any	approximately	are	as	at
be	because	been	before	being	between
both	but	by	can	could	did
do	does	done	due	during	each
either	enough	especially	etc	followed	following
for	found	from	further	give	given
giving	had	hardly	has	have	having
here	how	however	if	in	into
is	it	its	itself	just	kg
km	largely	like	made	mainly	make
may	might	min	ml	mm	more
most	mostly	must	nearly	neither	no
nor	not	now	obtain	obtained	of
often	on	only	or	other	our
out	over	overall	per	perhaps	possible
previously	quite	rather	really	regarding	resulted
resulting	same	seem	seen	several	should
show	showed	shown	shows	significant	significantly
since	so	some	such	suggest	than
that	the	their	theirs	them	then
there	these	they	this	those	through
thus	to	under	up	upon	use
used	using	various	very	was	we
were	what	when	whereas	which	while
with	within	without	would		

Appendix B. Query Results for Threshold

Table B.1: 20 Newsgroups Top 50 Results for *tire* LDAWN-QM without multiplier.

Rank ID	Doc ID	Score	Contains Query Term
1	871	0.03316	FALSE
2	804	0.01869	FALSE
3	1416	0.01591	FALSE
4	1289	0.01048	FALSE
5	1070	0.01043	FALSE
6	935	0.00971	FALSE
7	798	0.00931	FALSE
8	770	0.00915	FALSE
9	867	0.00895	FALSE
10	865	0.00849	FALSE
11	855	0.00824	FALSE
12	1247	0.00809	FALSE
13	1028	0.00785	TRUE
14	648	0.00708	FALSE
15	1237	0.00704	FALSE
16	1206	0.00691	FALSE
17	1322	0.00679	FALSE
18	1059	0.00669	FALSE
19	1327	0.00651	FALSE
20	889	0.0062459	FALSE
21	1417	0.006224	FALSE
22	1218	0.0061462	FALSE
23	969	0.0061358	FALSE
24	680	0.0058966	FALSE
25	879	0.0057491	FALSE
26	1305	0.0055068	FALSE
27	1167	0.0053424	FALSE
28	771	0.0053076	FALSE
29	1486	0.0052917	FALSE
30	1423	0.0052483	FALSE
31	1139	0.005112	FALSE
32	1242	0.0047165	FALSE
33	1296	0.0044527	FALSE
34	1379	0.0044466	FALSE
35	934	0.0043473	FALSE
36	717	0.0043185	FALSE
37	1278	0.004168	FALSE
38	932	0.0041277	FALSE
39	1311	0.0041182	FALSE
40	1270	0.0040925	FALSE
41	870	0.004047	FALSE
42	888	0.0039226	FALSE
43	992	0.0039171	FALSE
44	962	0.0038727	FALSE
45	723	0.0038614	FALSE
46	1152	0.0038458	FALSE
47	722	0.0037731	FALSE
48	650	0.0037452	FALSE
49	1119	0.0037276	FALSE
50	1261	0.0037211	FALSE

Table B.2:   20 Newsgroups Top 50 Results for *tire* LDAWN-QM with multiplier.

Rank ID	Doc ID	Score	Contains Query Term
1	871	0.033166	FALSE
2	804	0.018692	FALSE
3	1416	0.015908	FALSE
4	1289	0.010484	FALSE
5	1070	0.010436	FALSE
6	935	0.00971	FALSE
7	798	0.009301	FALSE
8	770	0.009154	FALSE
9	867	0.008951	FALSE
10	778	0.008578	TRUE
11	865	0.008495	FALSE
12	855	0.00824	FALSE
13	1247	0.008095	FALSE
14	1028	0.078541	TRUE
15	648	0.007084	FALSE
16	1237	0.007041	FALSE
17	1206	0.006914	FALSE
18	1322	0.006791	FALSE
19	1389	0.006758	TRUE
20	1059	0.006699	FALSE
21	1327	0.006519	FALSE
22	889	0.006246	FALSE
23	1417	0.006224	FALSE
24	1218	0.006146	FALSE
25	969	0.006136	FALSE
26	680	0.005897	FALSE
27	879	0.005749	FALSE
28	1305	0.005507	FALSE
29	1167	0.005342	FALSE
30	771	0.005308	FALSE
31	1486	0.005292	FALSE
32	1423	0.005248	FALSE
33	1139	0.005112	FALSE
34	1242	0.004716	FALSE
35	1296	0.004453	FALSE
36	1379	0.004447	FALSE
37	934	0.004347	FALSE
38	717	0.004318	FALSE
39	1278	0.004168	FALSE
40	932	0.004128	FALSE
41	1311	0.004118	FALSE
42	1270	0.004092	FALSE
43	870	0.004047	FALSE
44	888	0.003923	FALSE
45	992	0.003917	FALSE
46	962	0.003873	FALSE
47	723	0.003861	FALSE
48	1152	0.003846	FALSE
49	527	0.003801	TRUE
50	722	0.003773	FALSE

Table B.3:   20 Newsgroups 50 Results for *tire* LDAWN-QM that include threshold.

	Rank ID	Doc ID	Score	Contains Query Term
	120	255	0.001984976	TRUE
	121	773	0.001931834	FALSE
T	22	344	0.00181754	TRUE
	123	1271	0.001896472	FALSE
	124	998	0.001854854	FALSE
	125	1257	0.001853723	FALSE
	126	1306	0.001844183	FALSE
	127	909	0.001833978	FALSE
	128	1101	0.001757613	FALSE
	129	1362	0.001756074	FALSE
	130	653	0.001753169	FALSE
	131	953	0.001731787	FALSE
	132	1134	0.001727033	FALSE
	133	1404	0.001701206	FALSE
	134	1329	0.001664319	FALSE
	135	1316	0.001659924	FALSE
	136	813	0.00164674	FALSE
	136	272	0.001637991	FALSE
	137	380	0.001637991	FALSE
	138	1222	0.001597562	FALSE
	139	857	0.001554818	FALSE

Table B.4:   20 Newsgroups Top 50 Results for *ride* LDAWN-QM without multiplier.

Rank ID	Doc ID	Probability	Contains Query Term
1	871	0.186060441	FALSE
2	997	0.051509729	FALSE
3	1267	0.041896609	FALSE
4	659	0.038335711	FALSE
5	920	0.036671619	FALSE
6	1167	0.034976524	FALSE
7	906	0.031856524	FALSE
8	1104	0.031633455	FALSE
9	697	0.028778403	FALSE
10	1110	0.027113735	FALSE
11	1383	0.026841662	FALSE
12	650	0.02653452	FALSE
13	1096	0.022984068	TRUE
14	1154	0.021173054	FALSE
15	1413	0.02100343	FALSE
16	1366	0.020974831	FALSE
17	847	0.020647366	FALSE
18	799	0.020207448	FALSE
19	1249	0.019873187	FALSE
20	879	0.019868319	FALSE
21	1270	0.019835712	FALSE
22	1318	0.019395965	FALSE
23	1305	0.019191009	FALSE
24	1022	0.018470458	FALSE
25	1168	0.017953555	FALSE
26	1290	0.017291717	FALSE
27	1268	0.017245711	FALSE
28	1087	0.016661759	FALSE
29	1278	0.016418529	FALSE
30	820	0.016267066	TRUE
31	1099	0.01623675	FALSE
32	1122	0.016152947	FALSE
33	1007	0.016113107	FALSE
34	902	0.015946548	FALSE
35	771	0.015729584	FALSE
36	992	0.015299456	FALSE
37	1139	0.01519728	FALSE
38	1009	0.01506402	FALSE
39	1309	0.015029866	FALSE
40	1178	0.014952459	FALSE
41	1042	0.014946849	FALSE
42	977	0.014814756	FALSE
43	687	0.014484631	FALSE
44	776	0.014430382	FALSE
45	1308	0.013860994	FALSE
46	893	0.013802844	FALSE
47	972	0.013777189	FALSE
48	867	0.013555228	FALSE
49	1262	0.013416542	FALSE
50	842	0.013377748	FALSE

Table B.5: 20 Newsgroups Top 50 Results for *ride* LDAWN-QM with multiplier.

Rank ID	Doc ID	Probability	Contains Query Term
1	1096	0.229840681	TRUE
2	871	0.186060441	FALSE
3	820	0.162670664	TRUE
4	794	0.110861376	TRUE
5	816	0.097654596	TRUE
6	1256	0.092041269	TRUE
7	1225	0.074120896	TRUE
8	1430	0.071602667	TRUE
9	975	0.069736139	TRUE
10	980	0.060404707	TRUE
11	819	0.057359675	TRUE
12	997	0.051509729	FALSE
13	840	0.050928704	TRUE
14	961	0.046483136	TRUE
15	837	0.044719321	TRUE
16	821	0.042471929	TRUE
17	1267	0.041896609	FALSE
18	1319	0.040548607	TRUE
19	659	0.038335711	FALSE
20	920	0.036671619	FALSE
21	896	0.036026345	TRUE
22	1167	0.034976524	FALSE
23	994	0.034417739	TRUE
24	906	0.031856524	FALSE
25	1104	0.031633455	FALSE
26	982	0.030955756	TRUE
27	697	0.028778403	FALSE
28	1110	0.027113735	FALSE
29	1383	0.026841662	FALSE
30	650	0.02653452	FALSE
31	677	0.024933146	TRUE
32	1402	0.024258701	TRUE
33	698	0.022738164	TRUE
34	717	0.022579184	TRUE
35	1152	0.022497773	TRUE
36	1455	0.022435756	TRUE
37	584	0.022269541	TRUE
38	862	0.021938995	TRUE
39	656	0.021930943	TRUE
40	685	0.021325564	TRUE
41	1154	0.021173054	FALSE
42	522	0.021084543	TRUE
43	1413	0.02100343	FALSE
44	1366	0.020974831	FALSE
45	1474	0.020736291	TRUE
46	847	0.020647366	FALSE
47	799	0.020207448	FALSE
49	971	0.020199335	TRUE
50	1476	0.020053485	TRUE

Table B.6: 20 Newsgroups 50 Results for *ride* LDAWN-QM that include threshold.

Rank ID	Doc ID	Probability	Contains Query Term
1235	18	0.007073184	TRUE
1236	79	0.006986029	TRUE
1237	63	0.006582291	TRUE
1238	71	0.006555134	TRUE
1239	57	0.006500099	TRUE
T1240	98	0.006475024	TRUE
1241	12	0.006130599	FALSE
1242	82	0.004790998	FALSE
1243	32	0.004717674	FALSE
1244	75	0.004287192	FALSE
1245	77	0.003508338	FALSE
1246	11	0.003288017	FALSE
1247	68	0.002891067	FALSE
1248	59	0.002830257	FALSE
1249	92	0.002747409	FALSE
1250	66	0.002740194	FALSE
1251	10	0.002714924	FALSE
1252	80	0.002394469	FALSE
1253	52	0.0022163	FALSE
1254	89	0.002202194	FALSE
1255	81	0.002196211	FALSE
1256	36	0.002179058	FALSE
1257	96	0.001848448	FALSE
1258	14	0.001833192	FALSE
1259	53	0.001803524	FALSE

Table B.7:   20 Newsgroups Top 50 Results for *motorcycle* LDAWN-QM without multiplier.

Rank ID	Doc ID	Score	Contains Query Term
1	871	0.117176798	FALSE
2	804	0.062501071	FALSE
3	1416	0.052940055	FALSE
4	1289	0.036138009	FALSE
5	1070	0.035528231	FALSE
6	867	0.034383531	FALSE
7	935	0.033191297	FALSE
8	865	0.029586907	FALSE
9	798	0.028246779	FALSE
10	1247	0.028212869	FALSE
11	855	0.028010421	FALSE
12	770	0.027039503	FALSE
13	1237	0.024510269	FALSE
14	1206	0.023618162	FALSE
15	1059	0.022875663	FALSE
16	1327	0.022266261	FALSE
17	648	0.021653294	FALSE
18	879	0.021293224	FALSE
19	889	0.021180004	FALSE
20	680	0.019922668	FALSE
21	1028	0.019786048	FALSE
22	1167	0.019766919	FALSE
23	1218	0.018841765	FALSE
24	771	0.018705694	FALSE
25	1486	0.0178564	FALSE
26	1423	0.017592129	FALSE
27	1139	0.017447937	FALSE
28	1296	0.016641601	FALSE
29	1242	0.016095237	FALSE
30	932	0.015388416	FALSE
31	1379	0.01502117	FALSE
32	1311	0.013816661	FALSE
33	860	0.013751373	FALSE
34	1270	0.013751318	FALSE
35	870	0.013653339	FALSE
36	659	0.013514308	FALSE
37	888	0.013378781	FALSE
38	992	0.0133583	FALSE
39	1152	0.01296575	FALSE
40	1305	0.01294397	FALSE
41	723	0.012816795	FALSE
42	722	0.012800078	FALSE
43	650	0.012762889	FALSE
44	1119	0.012712224	FALSE
45	1261	0.012688215	FALSE
46	934	0.012686393	FALSE
47	1322	0.012563995	FALSE
48	1356	0.012346667	FALSE
49	1415	0.01294397	FALSE
50	256	0.012816795	FALSE

Table B.8:   20 Newsgroups Top 50 Results for *motorcycle* LDAWN-QM with multiplier.

Rank ID	Doc ID	Score	Contains Query Term
1	871	0.117176798	FALSE
2	1422	0.082752853	TRUE
3	804	0.062501071	FALSE
4	1416	0.052940055	FALSE
5	914	0.038032234	TRUE
6	1289	0.036138009	FALSE
7	1070	0.035528231	FALSE
8	867	0.034383531	FALSE
9	935	0.033191297	FALSE
10	865	0.029586907	FALSE
11	798	0.028246779	FALSE
12	1247	0.028212869	FALSE
13	855	0.028010421	FALSE
14	770	0.027039503	FALSE
15	1237	0.024510269	FALSE
16	1206	0.023618162	FALSE
17	1059	0.022875663	FALSE
18	1327	0.022266261	FALSE
19	648	0.021653294	FALSE
20	879	0.021293224	FALSE
21	889	0.021180004	FALSE
22	680	0.019922668	FALSE
23	1028	0.019786048	FALSE
24	1167	0.019766919	FALSE
25	1218	0.018841765	FALSE
26	771	0.018705694	FALSE
27	1486	0.0178564	FALSE
28	1423	0.017592129	FALSE
29	1139	0.017447937	FALSE
30	1296	0.016641601	FALSE
31	1242	0.016095237	FALSE
32	932	0.015388416	FALSE
33	1379	0.01502117	FALSE
34	559	0.014282359	TRUE
35	1276	0.014232525	TRUE
36	661	0.014222617	TRUE
37	933	0.014222617	TRUE
38	1032	0.01419301	TRUE
39	1451	0.01419301	TRUE
40	958	0.01419301	TRUE
41	1193	0.014173368	TRUE
42	565	0.014144049	TRUE
43	793	0.014134314	TRUE
44	121	0.014134314	TRUE
45	1157	0.014124598	TRUE
46	332	0.014124598	TRUE
47	689	0.014124598	TRUE
48	1109	0.014095564	TRUE
49	986	0.014057114	TRUE
50	4	0.014038	TRUE

81

Table B.9: 20 Newsgroups 50 Results for *motorcycle* LDAWN-QM that include threshold.

Rank ID	Doc ID	Score	Contains Query Term
265	276	0.00457768	FALSE
T 266	1440	0.00456982	TRUE
267	1444	0.00454923	FALSE
268	719	0.00452278	FALSE
269	961	0.00450029	FALSE
270	835	0.00449955	FALSE
271	716	0.0044845	FALSE
272	704	0.00448392	FALSE
273	1072	0.00447425	FALSE
274	229	0.00445212	FALSE
275	1064	0.00445026	FALSE
276	1354	0.00444029	FALSE
277	653	0.00441557	FALSE
278	1324	0.00439023	FALSE
279	1373	0.00439011	FALSE
280	837	0.00438833	FALSE
281	1052	0.00438476	FALSE
282	1186	0.00437225	FALSE
283	1243	0.00432614	FALSE
285	621	0.00428369	FALSE
286	820	0.00426182	FALSE
287	1294	0.00419063	FALSE
288	1362	0.0041457	FALSE
289	1150	0.00414112	FALSE
290	1129	0.00413235	FALSE
291	951	0.00412867	FALSE
292	1132	0.00412767	FALSE
293	853	0.00411506	FALSE
294	688	0.00410878	FALSE
295	1331	0.00410722	FALSE
296	273	0.00408832	FALSE
297	1375	0.00408512	FALSE
298	640	0.00408117	FALSE
299	1240	0.00408016	FALSE
300	1138	0.00407834	FALSE
301	359	0.00407813	FALSE
302	1016	0.00407782	FALSE
303	651	0.00407072	FALSE
304	726	0.00405278	FALSE
305	1096	0.00393794	FALSE
306	70	0.00390375	FALSE
307	906	0.00390125	FALSE
308	88	0.00389169	FALSE
309	326	0.00388987	FALSE
400	963	0.00385878	FALSE
401	1205	0.00383286	FALSE
402	756	0.00382452	FALSE
403	1082	0.00381066	FALSE
404	1182	0.00380587	FALSE
405	918	0.00380058	FALSE

Table B.10: 20 Newsgroups Top 50 Results for *battery* LDAWN-QM without multiplier.

Rank ID	Doc ID	Score	Contains Query Term
1	1488	0.052216124	FALSE
2	252	0.020199892	FALSE
3	150	0.019438001	FALSE
4	1444	0.014386422	FALSE
5	300	0.009542432	FALSE
6	472	0.009292006	FALSE
7	62	0.008992184	FALSE
8	65	0.007724998	FALSE
9	1468	0.007003791	FALSE
10	1247	0.006582145	FALSE
11	1453	0.006465551	FALSE
12	803	0.006285569	FALSE
13	1176	0.006186287	FALSE
14	348	0.006094796	FALSE
15	1229	0.006078649	FALSE
16	113	0.005716533	FALSE
17	245	0.005548136	FALSE
18	1141	0.005522173	FALSE
19	1470	0.005095251	FALSE
20	242	0.004940127	FALSE
21	1490	0.004895259	FALSE
22	17	0.004875446	FALSE
23	525	0.004834639	FALSE
24	484	0.004722107	FALSE
25	612	0.004502292	FALSE
26	1477	0.004497616	FALSE
27	437	0.00442602	FALSE
28	830	0.004418234	FALSE
29	1186	0.004301852	FALSE
30	1475	0.004281498	FALSE
31	1437	0.004266579	FALSE
32	593	0.004258088	FALSE
33	1473	0.004246195	FALSE
34	187	0.004204766	FALSE
35	35	0.003992296	FALSE
36	648	0.003974289	FALSE
37	531	0.003959037	FALSE
38	1069	0.00391493	FALSE
39	12	0.003835613	FALSE
40	537	0.003770945	FALSE
41	406	0.003740028	FALSE
42	375	0.003694628	FALSE
43	1411	0.003645722	FALSE
44	228	0.003547504	FALSE
45	596	0.003546984	FALSE
46	1441	0.003470017	FALSE
47	305	0.003389613	FALSE
48	134	0.003330222	FALSE
49	174	0.00324128	FALSE
50	516	0.003112821	FALSE

Table B.11: 20 Newsgroups Top 50 Results for *battery* LDAWN-QM with multiplier.

Rank ID	Doc ID	Score	Contains Query Term
1	1488	0.052216124	FALSE
2	1034	0.026558529	TRUE
3	252	0.020199892	FALSE
4	150	0.019438001	FALSE
5	687	0.019206325	TRUE
6	876	0.017169668	TRUE
7	1444	0.014386422	FALSE
8	987	0.012699838	TRUE
9	1235	0.01218218	TRUE
10	508	0.01203805	TRUE
11	640	0.011251489	TRUE
12	180	0.010081401	TRUE
13	894	0.010030106	TRUE
14	300	0.009542432	FALSE
15	472	0.009292006	FALSE
16	2	0.0092181	TRUE
17	62	0.008992184	FALSE
18	713	0.008852188	TRUE
19	1100	0.00875712	TRUE
20	75	0.008132286	TRUE
21	226	0.008052981	TRUE
22	65	0.007724998	FALSE
23	768	0.007713826	TRUE
24	14	0.007584395	TRUE
25	532	0.007294821	TRUE
26	592	0.007185885	TRUE
27	1468	0.007003791	FALSE
28	3	0.006731796	TRUE
29	1247	0.006582145	FALSE
30	1453	0.006465551	FALSE
31	281	0.00630454	TRUE
32	803	0.006285569	FALSE
33	1420	0.00624071	TRUE
34	1176	0.006186287	FALSE
35	348	0.006094796	FALSE
36	41	0.006085924	TRUE
37	1229	0.006078649	FALSE
38	1319	0.005877533	TRUE
39	113	0.005716533	FALSE
40	245	0.005548136	FALSE
41	1141	0.005522173	FALSE
42	1470	0.005095251	FALSE
43	242	0.004940127	FALSE
44	1490	0.004895259	FALSE
45	17	0.004875446	FALSE
46	525	0.004834639	FALSE
47	484	0.004722107	FALSE
48	402	0.004720664	TRUE
49	415	0.004649911	TRUE
50	612	0.004502292	FALSE

Table B.12:   20 Newsgroups Results for *battery* LDAWN-QM that include threshold.

	Rank ID	Doc ID	Score	Contains Query Term
	215	910	0.001474952	FALSE
	216	551	0.001472451	FALSE
	217	638	0.001464582	FALSE
	218	137	0.001463649	FALSE
	219	94	0.001455787	FALSE
	220	1449	0.001455787	FALSE
	221	21	0.001453438	FALSE
	222	862	0.001444799	FALSE
	223	1184	0.001440893	FALSE
	224	663	0.001433002	FALSE
	225	512	0.001415727	FALSE
	226	643	0.001402661	FALSE
	227	31	0.001401204	FALSE
	228	260	0.001388278	FALSE
	229	554	0.001387688	FALSE
	230	466	0.001387688	FALSE
	231	624	0.001381635	FALSE
	232	256	0.001380915	FALSE
	233	1044	0.001380908	FALSE
	234	315	0.001380908	FALSE
	235	145	0.001380727	FALSE
	236	1077	0.001380008	FALSE
	237	1194	0.001379829	FALSE
	238	401	0.001377539	FALSE
T	239	383	0.001377408	TRUE
	240	1266	0.001377365	FALSE
	241	461	0.001371682	FALSE
	242	278	0.001369303	FALSE
	243	581	0.001366545	FALSE
	244	489	0.001363311	FALSE

Table B.13: NIPS Results for *Bayes* LDAWN-QM without multiplier.

Rank ID	Doc ID	Score	Contains Query Term
1	207	0.61727948	FALSE
2	120	0.330365769	FALSE
3	247	0.28132891	FALSE
4	345	0.180628114	FALSE
5	337	0.174045685	FALSE
6	223	0.17222019	TRUE
7	277	0.171644593	FALSE
8	114	1.655974304	TRUE
9	26	0.162346557	FALSE
10	167	0.1614067	FALSE
11	150	0.15654238	FALSE
12	308	0.15524306	FALSE
13	213	0.154092282	FALSE
14	271	0.152999172	FALSE
15	281	0.147494736	FALSE
16	149	0.147031095	TRUE
17	348	0.140448695	FALSE
18	50	0.135807865	FALSE
19	392	0.122825616	FALSE
20	27	0.122757842	FALSE
21	356	0.118059607	FALSE
22	166	0.115141304	FALSE
23	239	0.112123961	FALSE
24	28	0.10925041	FALSE
25	134	0.107825884	FALSE
26	230	0.107377274	FALSE
27	262	0.1071092	FALSE
28	41	0.105333859	FALSE
29	83	0.105005367	FALSE
30	169	0.099350174	FALSE
31	242	0.093183323	TRUE
32	322	0.092966828	FALSE
33	228	0.086878106	FALSE
34	137	0.08487155	TRUE
35	139	0.082242785	TRUE
36	288	0.08136031	FALSE
37	47	0.079483263	FALSE
38	254	0.078420086	TRUE
39	165	0.078383783	FALSE
40	62	0.77848299	TRUE
41	37	0.077668305	FALSE
42	265	0.077598796	TRUE
43	94	0.075952491	FALSE
44	59	0.074388732	FALSE
45	15	0.06914799	FALSE
46	355	0.069028807	FALSE
47	226	0.068803801	FALSE
48	284	0.067756393	FALSE
49	244	0.067367185	FALSE
50	257	0.065738686	FALSE

Table B.14:   NIPS Results for *Bayes* LDAWN-QM with multiplier.

Rank ID	Doc ID	Score	Contains Query Term
1	223	1.722201903	TRUE
2	114	1.655974304	TRUE
3	149	1.470310947	TRUE
4	242	0.931833226	TRUE
5	137	0.848715499	TRUE
6	139	0.822427849	TRUE
7	254	0.784200862	TRUE
8	62	0.77848299	TRUE
9	265	0.775987958	TRUE
10	365	0.654048207	TRUE
11	399	0.637457882	TRUE
12	207	0.61727948	FALSE
13	298	0.572386389	TRUE
14	54	0.568542598	TRUE
15	371	0.470205358	TRUE
16	377	0.445110273	TRUE
17	282	0.437756748	TRUE
18	341	0.434836692	TRUE
19	342	0.431972855	TRUE
20	258	0.418378326	TRUE
21	157	0.397198316	TRUE
22	78	0.396897861	TRUE
23	145	0.38507426	TRUE
24	340	0.362230288	TRUE
25	120	0.330365769	FALSE
26	306	0.31090788	TRUE
27	231	0.290864431	TRUE
28	247	0.28132891	FALSE
29	283	0.27179743	TRUE
30	75	0.255307912	TRUE
31	199	0.222797276	TRUE
32	44	0.222557846	TRUE
33	35	0.218302563	TRUE
34	21	0.208848634	TRUE
35	374	0.206282976	TRUE
36	195	0.194420026	TRUE
37	80	0.194282521	TRUE
38	30	0.187062398	TRUE
39	370	0.183797186	TRUE
40	318	0.181370208	TRUE
41	345	0.180628114	FALSE
42	102	0.180071478	TRUE
43	337	0.174045685	FALSE
44	277	0.171644593	FALSE
45	29	0.171381588	TRUE
46	7	0.168093522	TRUE
47	369	0.162437845	TRUE
48	26	0.162346557	FALSE
49	167	0.1614067	FALSE
50	150	0.15654238	FALSE

Table B.15:   NIPS Results for *Bayes* LDAWN-QM that include threshold.

Rank ID	Doc ID	Score	Contains Query Term
246	63	0.018220736	FALSE
247	194	0.0181195	TRUE
248	6	0.01798103	FALSE
249	323	0.017928973	TRUE
250	314	0.017909591	FALSE
251	136	0.017814952	FALSE
252	315	0.017428103	FALSE
253	32	0.017285932	FALSE
254	110	0.017018882	FALSE
255	168	0.016942715	FALSE
256	245	0.016841418	FALSE
257	42	0.016781478	FALSE
258	327	0.016625765	FALSE
259	152	0.016519702	FALSE
260	196	0.016287481	FALSE
261	172	0.016248705	FALSE
262	330	0.016030835	FALSE
263	347	0.015977615	FALSE
264	25	0.01597702	FALSE
265	17	0.01595203	FALSE
266	177	0.015947622	FALSE
267	65	0.015946216	FALSE
268	159	0.015176747	FALSE
269	375	0.014957044	FALSE
270	131	0.014950257	FALSE
271	198	0.014906954	FALSE
272	316	0.014895864	FALSE
273	97	0.014882671	FALSE
274	51	0.014852667	FALSE
275	291	0.014810842	FALSE
276	45	0.014419779	FALSE
277	390	0.014391904	FALSE
278	61	0.014240917	FALSE
279	217	0.01417662	FALSE
280	68	0.014062357	FALSE
281	214	0.014022284	TRUE
282	171	0.014008986	FALSE
283	191	0.014000301	FALSE
284	313	0.01393958	FALSE
285	225	0.013884601	FALSE

Table B.16:   OHSUMED Results for *cells* LDAWN-QM without multiplier.

Rank ID	Doc ID	Score	Contains Query Term
1	16042	0.9232724	FALSE
2	13287	0.9169833	FALSE
3	16481	0.871207	FALSE
4	6783	0.8301589	FALSE
5	16783	0.8031941	FALSE
6	16463	0.7989352	FALSE
7	13717	0.7937039	TRUE
8	10561	0.7806776	TRUE
9	15553	0.7475219	FALSE
10	10423	0.7303034	TRUE
11	2596	0.7212073	FALSE
12	13484	0.7063251	FALSE
13	12960	0.6976543	FALSE
14	11964	0.6966956	TRUE
15	308	0.6881246	TRUE
16	21	0.6810932	FALSE
17	15209	0.6724413	TRUE
18	332	0.6710078	TRUE
19	12771	0.6648887	TRUE
20	3366	0.6640888	TRUE
21	306	0.6622016	TRUE
22	4169	0.6447832	TRUE
23	16607	0.6447377	TRUE
24	16859	0.6683586	TRUE
25	20	0.6359893	TRUE
26	1395	0.6334216	TRUE
27	609	0.6314459	TRUE
28	11981	0.6229398	TRUE
29	15486	0.6227939	TRUE
30	10239	0.6224376	FALSE
31	10240	0.616607	TRUE
32	11772	0.6100517	TRUE
33	12105	0.6003751	FALSE
34	2278	0.598789	TRUE
35	10769	0.5975317	FALSE
36	5686	0.5861926	TRUE
37	4431	0.5671678	TRUE
38	13786	0.5649877	TRUE
39	305	0.5648526	TRUE
40	13957	0.5616534	TRUE
41	8569	0.5561922	TRUE
42	6685	0.5536857	TRUE
43	10803	0.5531038	TRUE
44	9935	0.5507056	FALSE
45	1399	0.5437042	FALSE
46	2589	0.542408	FALSE
47	17058	0.5393018	FALSE
48	10420	0.5383141	TRUE
49	8583	0.5364693	TRUE
50	13299	0.5346369	FALSE

Table B.17: OHSUMED Results for *cells* LDAWN-QM with multiplier.

Rank ID	Doc ID	Score	Contains Query Term
1	13717	0.793704	TRUE
2	10561	0.780678	TRUE
3	10423	0.730303	TRUE
4	11964	0.696696	TRUE
5	308	0.688125	TRUE
6	15209	0.672441	TRUE
7	332	0.671008	TRUE
8	12771	0.664889	TRUE
9	3366	0.664089	TRUE
10	306	0.662202	TRUE
11	4169	0.644783	TRUE
12	16607	0.644738	TRUE
13	16859	0.638359	TRUE
14	20	0.635989	TRUE
15	1395	0.633422	TRUE
16	609	0.631446	TRUE
17	11981	0.62294	TRUE
18	15486	0.622794	TRUE
19	10240	0.616607	TRUE
20	11772	0.610052	TRUE
21	2278	0.598789	TRUE
22	5686	0.586193	TRUE
23	4431	0.561678	TRUE
24	13786	0.564988	TRUE
25	305	0.564853	TRUE
26	13957	0.561653	TRUE
27	8569	0.556192	TRUE
28	6685	0.553686	TRUE
29	10803	0.553104	TRUE
30	10420	0.538314	TRUE
31	8583	0.536469	TRUE
32	16478	0.528398	TRUE
33	15619	0.522918	TRUE
34	8328	0.515919	TRUE
35	9670	0.511905	TRUE
36	282	0.510285	TRUE
37	7150	0.506628	TRUE
38	9344	0.506447	TRUE
39	15087	0.505576	TRUE
40	2208	0.502285	TRUE
41	11389	0.49278	TRUE
42	15488	0.490816	TRUE
43	7798	0.487731	TRUE
44	334	0.486958	TRUE
45	15512	0.482224	TRUE
46	5329	0.471647	TRUE
47	6691	0.468509	TRUE
48	9342	0.467987	TRUE
49	16036	0.463657	TRUE
50	15866	0.463262	TRUE

Table B.18:   OHSUMED Results for *cells* LDAWN-QM with threshold.

Rank ID	Doc ID	Score	Contains Query Term
8294	15465	0.129142944	TRUE
8295	11278	0.129141143	TRUE
8296	944	0.129140428	TRUE
8297	14016	0.12913936	TRUE
8298	11328	0.129139006	TRUE
8299	1785	0.129138652	TRUE
8300	12947	0.129136894	TRUE
8301	3763	0.129135501	TRUE
8302	680	0.129134118	TRUE
8303	2189	0.129132405	TRUE
8304	8159	0.129132064	TRUE
8305	3114	0.129132064	TRUE
T 8306	11064	0.129122794	TRUE
8307	5183	0.129030825	FALSE
8308	14567	0.12896494	FALSE
8309	13351	0.128947968	FALSE
8310	4371	0.128600623	FALSE
8311	14537	0.128376079	FALSE
8312	4227	0.128203829	FALSE
8313	11824	0.128107405	FALSE
8314	14020	0.128039241	FALSE
8315	832	0.127879954	FALSE
8316	4952	0.127791261	FALSE
8317	2022	0.127780221	FALSE
8318	3342	0.12746014	FALSE
8319	14924	0.127412859	FALSE
8320	13092	0.12715341	FALSE
8321	16081	0.127081281	FALSE
8322	14091	0.127044827	FALSE
8323	6816	0.127022405	FALSE
8324	11504	0.127018784	FALSE
8325	15670	0.126961653	FALSE
8326	9099	0.12692029	FALSE

Bibliography

1. *Random House Webster's unabridged dictionary.* Random House, New York, 2. ed., [nachdr.] edition, 2004,2001. ISBN 9780375425998.

2. Alpert, Jesse and Nissan Hajaj. "We knew the web was big..." 991–1000. ACM, New York, NY, USA, 2009. ISBN 978-1-60558-487-4.

3. Antoniou, Grigoris and Frank Van Harmelen. "Web Ontology Language: OWL". *Handbook on Ontologies in Information Systems*, 67–92. Springer, 2003.

4. Baezo-Yates, Ricardo and Berthier Ribero-Neto. *Modern Information Retrieval.* Addison-Wesley, 1999.

5. Baker, L. Douglas and Andrew Kachites McCallum. "Distributional clustering of words for text classification". *SIGIR '98: Proceedings of the 21st annual international ACM SIGIR conference on Research and development in information retrieval*, 96–103. ACM, New York, NY, USA, 1998. ISBN 1-58113-015-5.

6. Blanco, Roi and Christina Lioma. "Random walk term weighting for information retrieval". *SIGIR '07: Proceedings of the 30th annual international ACM SIGIR conference on Research and development in information retrieval*, 829–830. ACM, New York, NY, USA, 2007. ISBN 978-1-59593-597-7.

7. Blei, David M., Andrew Y. Ng, and Michael I. Jordan. "Latent dirichlet allocation". *J. Mach. Learn. Res.*, 3:993–1022, 2003. ISSN 1533-7928.

8. Buckley, Chris and Gerard Salton. "Stopword List 2". http://www.lextek.com/manuals/onix/stopwords2.html. [Online; accessed 28-June-2009].

9. Budanitsky, Alexander and Graeme Hirst. "Evaluating WordNet-based Measures of Lexical Semantic Relatedness". *Comput. Linguist.*, 32(1):13–47, 2006. ISSN 0891-2017.

10. Chechik, G. "NIPS". Online. http://ai.stanford.edu/gal., 2008.

11. Chua, Stephanie and Narayanan Kulathuramaiyer. "Semantic Feature Selection Using WordNet". *WI '04: Proceedings of the 2004 IEEE/WIC/ACM International Conference on Web Intelligence*, 166–172. IEEE Computer Society, Washington, DC, USA, 2004. ISBN 0-7695-2100-2.

12. Fresno, Víctor, Raquel Martínez, and Soto Montalvo. "Improving Web Page Clustering Through Selecting Appropriate Term Weighting Functions". *ICDIM*, 511–518. 2006.

13. Garcia, Dr. E. "Mi lslita on-line Tutorial". URL http://www.miislita.com/information-retrieval-tutorial. [Online; accessed 20-July-2009].

14. Globerson, Amir, Gal Chechik, Fernando Pereira, and Naftali Tishby. "Euclidean Embedding of Co-occurrence Data". 2007. URL http://ai.stanford.edu/ gal/data.html.

15. Griffiths, T.L. and M. Steyvers. "Finding scientific topics". 5228–5235. National Academy of Sciences, 2004.

16. Griffiths, T.L., M. Steyvers, and Jb Tenenbaum. "Topics in semantic representation." *Psychological Review*, 114(2):211–244, 2007.

17. He, Xiaofei, Deng Cai, Haifeng Liu, and Wei-Ying Ma. "Locality preserving indexing for document representation". *SIGIR '04: Proceedings of the 27th annual international ACM SIGIR conference on Research and development in information retrieval*, 96–103. ACM, New York, NY, USA, 2004. ISBN 1-58113-881-4.

18. Hearst, A., Marti. "Automated discovery of WordNet relations". *C. Fellbaum, WordNet: An Electronic Lexical Database*, 131–153. MIT Press, 1998. URL http://www.sims.berkeley.edu/ hearst/papers/wordnet98.pdf.

19. Hersh, William. "OHSUMED: an interactive retrieval evaluation and new large test collection for research". *SIGIR '94: Proceedings of the 17th annual international ACM SIGIR conference on Research and development in information retrieval*, 192–201. Springer-Verlag New York, Inc., New York, NY, USA, 1994. ISBN 0-387-19889-X.

20. Hofmann, Thomas. "Probabilistic Latent Semantic Analysis". *In Proc. of Uncertainty in Artificial Intelligence, UAI99*, 289–296. 1999.

21. Hotho, Andreas, Steffen Staab, and Gerd Stumme. "Wordnet improves Text Document Clustering". *In Proc. of the SIGIR 2003 Semantic Web Workshop*, 541–544. 2003.

22. How, Bong Chih, Narayanan Kulathuramaiyer, and Wong Ting Kiong. "Categorical Term Descriptor: A Proposed Term Weighting Scheme for Feature Selection". *WI '05: Proceedings of the 2005 IEEE/WIC/ACM International Conference on Web Intelligence*, 313–316. IEEE Computer Society, Washington, DC, USA, 2005. ISBN 0-7695-2415-X.

23. Jiang Yunfei, Guo Ping, Chen Xinyu and Lu Hanging. "An Improved Random Sampling LDA for Face Recognition". *CISP '08: Proceedings of the 2008 Congress on Image and Signal Processing, Vol. 2*, 685–689. IEEE Computer Society, Washington, DC, USA, 2008. ISBN 978-0-7695-3119-9.

24. Jiménez, D., E. Ferretti, V. Vidal, P. Rosso, and C. F. Enguix. "The influence of semantics in IR using LSI and K-means clustering techniques". *ISICT '03: Proceedings of the 1st international symposium on Information and communication technologies*, 279–284. Trinity College Dublin, 2003.

25. Jurafsky, Daniel and H. Martin, James. *Speech and language processing: an introduction to natural language*. Prentice Hall, 2008.

26. Kohonen, T. *Self-Organizing Maps*. New York: Springer, 3rd edition, 2001.

27. Kwok, K. L. "A new method of weighting query terms for ad-hoc retrieval". *SIGIR '96: Proceedings of the 19th annual international ACM SIGIR conference on Research and development in information retrieval*, 187–195. ACM, New York, NY, USA, 1996. ISBN 0-89791-792-8.

28. Lagus, Krista, Timo Honkela, Samuel Kaski, and Teuvo Kohonen. "WEBSOM for Textual Data Mining". *Artificial Intelligence Review*, 13:345–364, 1999.

29. Lang, K. "20 Newsgroups". Online. http://people.csail.mit.edu/jrennie/20Newsgroups/, 2008.

30. Liu, Shuang, Fang Liu, Clement Yu, and Weiyi Meng. "An effective approach to document retrieval via utilizing WordNet and recognizing phrases". *SIGIR '04: Proceedings of the 27th annual international ACM SIGIR conference on Research and development in information retrieval*, 266–272. ACM, New York, NY, USA, 2004. ISBN 1-58113-881-4.

31. Liu, Xiaoyong and W. Bruce Croft. "Cluster-based retrieval using language models". *SIGIR '04: Proceedings of the 27th annual international ACM SIGIR conference on Research and development in information retrieval*, 186–193. ACM, New York, NY, USA, 2004. ISBN 1-58113-881-4.

32. Lovins, J. B. "Development of a stemming algorithm". *Mechanical Translation and Computational Linguistics*, 11(1/2):22–31, 1968.

33. Mandala, Rila, Takenobu Tokunaga, and Hozumi Tokunaga. "Complementing WordNet with Roget's and corpus-based thesauri for information retrieval". *Proceedings of the ninth conference on European chapter of the Association for Computational Linguistics*, 94–101. Association for Computational Linguistics, Morristown, NJ, USA, 1999.

34. Manning, Christopher D., Prabhakar Raghavan, and Hinrich Schütze. *Introduction to Information Retrieval*. Cambridge University Press, July 2008. ISBN 0521865719.

35. Mansuy, Trevor and Robert J. Hilderman. "A characterization of wordNet features in Boolean models for text classification". *AusDM '06: Proceedings of the fifth Australasian conference on Data mining and analytics*, 103–109. Australian Computer Society, Inc., Darlinghurst, Australia, 2006. ISBN 1-920682-41-4.

36. Millar, Jeremy, Gilbert Peterson, and Michael Mendenhall. "Document Clustering and Visualization with Latent Dirichlet Allocation and Self-Organizing Maps", 2009. URL http://www.aaai.org/ocs/index.php/FLAIRS/2009/paper/view/62.

37. Miller, George A. "WordNet: a lexical database for English". *Commun. ACM*, 38(11):39–41, 1995. ISSN 0001-0782.

38. Nanas, Nikolaos, Victoria Uren, and Anne de Roeck. "A Comparative Evaluation of Term Weighting Methods for Information Filtering". *DEXA '04: Proceedings of the Database and Expert Systems Applications, 15th International Workshop*, 13–17. IEEE Computer Society, Wash, DC, USA, 2004. ISBN 0-7695-2195-9.

39. Pereira, Fernando, Naftali Tishby, and Lillian Lee. "Distributional clustering of English words". *Proceedings of the 31st annual meeting on Association for Computational Linguistics*, 183–190. Association for Computational Linguistics, Morristown, NJ, USA, 1993.

40. Reynaud, Chantal and Brigitte Safar. "Exploiting WordNet as Background Knowledge". *OM*. 2007.

41. Steyvers, M. and T. L. Griffiths. *Probabilistic topic models*. Laurence Erlbaum Associates, 2007.

42. Teich, Elke and Peter Fankhauser. "WordNet for Lexical Cohesion Analysis". *In Second International WordNet Conference*, 326–331. 2004.

43. University, Ted Pedersen, Ted Pedersen, and Siddharth Patwardhan. "Word-Net::Similarity - Measuring the Relatedness of Concepts". 1024–1025. 2004.

44. Varelas, Giannis, Epimenidis Voutsakis, Paraskevi Raftopoulou, Euripides G.M. Petrakis, and Evangelos E. Milios. "Semantic similarity methods in WordNet and their application to information retrieval on the web". *WIDM '05: Proceedings of the 7th annual ACM international workshop on Web information and data management*, 10–16. ACM, New York, NY, USA, 2005. ISBN 1-59593-194-5.

45. Voorhees, Ellen. "The TREC Conferences: An Introduction", 2008.

46. Wallace, Mike. "Jawbone Java WordNet API". Online. http://wordnet.princeton.edu/, 2007.

47. Wallach, Hanna M. "Topic modeling: beyond bag-of-words". *ICML '06: Proceedings of the 23rd international conference on Machine learning*, 977–984. ACM, New York, NY, USA, 2006. ISBN 1-59593-383-2.

48. Wei, Xing and W. Bruce Croft. "LDA-based document models for ad-hoc retrieval". *SIGIR '06: Proceedings of the 29th annual international ACM SIGIR conference on Research and development in information retrieval*, 178–185. ACM, New York, NY, USA, 2006. ISBN 1-59593-369-7.

49. Wu, Harry and Gerard Salton. "A comparison of search term weighting: term relevance vs. inverse document frequency". *SIGIR Forum*, 16(1):30–39, 1981. ISSN 0163-5840.

50. Xu, Hongzhi and Chunping Li. "A Novel Term Weighting Scheme for Automated Text Categorization". *ISDA '07: Proceedings of the Seventh International Conference on Intelligent Systems Design and Applications*, 759–764. IEEE Computer Society, Washington, DC, USA, 2007. ISBN 0-7695-2976-3.

51. Yue-Heng Sun, Zhi-Gang Chen, Pi-Lian He. "An improved term weighting scheme for vector space model". *ICML '04: Proceedings of the 3rd international conference on Machine learning and Cybernetics*, 1692–1695. IEEE Computer Society, Washington, DC, USA, 2004. ISBN 0-7803-8403-2.